RESEARCH METHODOLOGIES AND PRACTICAL APPLICATIONS OF CHEMISTRY

Innovations in Physical Chemistry: Monograph Series

RESEARCH METHODOLOGIES AND PRACTICAL APPLICATIONS OF CHEMISTRY

Edited by
Lionello Pogliani, PhD
A. K. Haghi, PhD
Nazmul Islam, PhD

APPLE
ACADEMIC
PRESS

Apple Academic Press Inc.	Apple Academic Press Inc.
3333 Mistwell Crescent	1265 Goldenrod Circle NE
Oakville, ON L6L 0A2	Palm Bay, Florida 32905
Canada USA	USA

ISBN 13: 978-1-77463-458-5 (pbk)
ISBN 13: 978-1-77188-784-7 (hbk)

Library and Archives Canada Cataloguing in Publication

Title: Research methodologies and practical applications of chemistry / edited by Lionello Pogliani, PhD, A.K. Haghi, PhD, Nazmul Islam, PhD.

Names: Pogliani, Lionello, editor. | Haghi, A. K., editor. | Islam, Nazmul, editor.

Series: Innovations in physical chemistry.

Description: Series statement: Innovations in physical chemistry | Includes bibliographical references and index.

Identifiers: Canadiana (print) 20190130725 | Canadiana (ebook) 20190130768 | ISBN 9781771887847 (hardcover) | ISBN 9780429023460 (ebook)

Subjects: LCSH: Chemistry, Technical.

Classification: LCC TP145 R47 2019 | DDC 660—dc23

CIP data on file with US Library of Congress

ABOUT THE EDITORS

Lionello Pogliani, PhD
University of Valencia-Burjassot, Spain
E-mail: lionello.pogliani@uv.es

Lionello Pogliani, PhD, is a retired professor of physical chemistry. He received his postdoctoral training at the Department of Molecular Biology of the C. E. A. (Centre d'Etudes Atomiques) of Saclay, France, at the Physical Chemistry Institute of the Technical and Free University of Berlin, and at the Pharmaceutical Department of the University of California, San Francisco, USA. He spent his sabbatical years at the Technical University of Lisbon, Portugal, and at the University of Valencia, Spain. He has contributed more than 200 papers in the experimental, theoretical, and didactical fields of physical chemistry, including chapters in specialized books and a book on numbers 0, 1, 2, and 3. A work of his has been awarded with the GM Neural Trauma Research Award. He is a member of the International Academy of Mathematical Chemistry and he is on the editorial board of many international journals. He is presently a part-time teammate at the Physical Chemistry Department of the University of Valencia, Spain.

A. K. Haghi, PhD
Professor Emeritus of Engineering Sciences, Former Editor-in-Chief, International Journal of Chemoinformatics and Chemical Engineering and Polymers Research Journal; Member, Canadian Research and Development Center of Sciences and Cultures (CRDCSC), Canada

A. K. Haghi, PhD, is the author and editor of 165 books, as well as 1000 published papers in various journals and conference proceedings. Dr. Haghi has received several grants, consulted for a number of major corporations, and is a frequent speaker to national and international audiences. Since 1983, he served as professor at several universities. He was formerly the editor-in-chief of the *International Journal of Chemoinformatics and Chemical Engineering* and *Polymers Research Journal* and on the editorial boards of many international journals. He is also a member of the Canadian Research and Development Center of Sciences and Cultures (CRDCSC), Montreal, Quebec, Canada.

Nazmul Islam, PhD
Professor, Department of Basic Sciences and Humanities, Techno Global-Balurghat, Balurghat, D. Dinajpur, India

Nazmul Islam, PhD, is now working as an assistant professor in the Department of Basic Sciences and Humanities at Techno Global-Balurghat (now Techno India-Balurghat), Balurghat, D. Dinajpur, India. He has published more than 60 research papers in several prestigious peer-reviewed journals and has written many book chapters and research books. In addition, he is the editor-in-chief of *The SciTech, Journal of Science and Technology, The SciTech, International Journal of Engineering Sciences*, and *The Signpost Open Access Journal of Theoretical Sciences*. He also serves as a member on the editorial boards of several journals.

Dr. Islam's research interests are in theoretical chemistry, particularly quantum chemistry, conceptual density functional theory (CDFT), periodicity, SAR, QSAR/QSPR study, drug design, HMO theory, biological function of chemical compounds, quantum biology, nanochemistry, and more.

INNOVATIONS IN PHYSICAL CHEMISTRY: MONOGRAPH SERIES

This book series offers a comprehensive collection of books on physical principles and mathematical techniques for majors, non-majors, and chemical engineers. Because there are many exciting new areas of research involving computational chemistry, nanomaterials, smart materials, high-performance materials, and applications of the recently discovered graphene, there can be no doubt that physical chemistry is a vitally important field. Physical chemistry is considered a daunting branch of chemistry—it is grounded in physics and mathematics and draws on quantum mechanics, thermodynamics, and statistical thermodynamics.

Editors-in-Chief

A. K. Haghi, PhD
Editor-in-Chief, *International Journal of Chemoinformatics* and *Chemical Engineering and Polymers Research Journal*; Member, Canadian Research and Development Center of Sciences and Cultures (CRDCSC), Montreal, Quebec, Canada
E-mail: AKHaghi@Yahoo.com

Lionello Pogliani, PhD
University of Valencia-Burjassot, Spain
E-mail: lionello.pogliani@uv.es

Ana Cristina Faria Ribeiro, PhD
Researcher, Department of Chemistry, University of Coimbra, Portugal
E-mail: anacfrib@ci.uc.pt

BOOKS IN THE SERIES

- **Applied Physical Chemistry with Multidisciplinary Approaches**
  Editors: A. K. Haghi, PhD, Devrim Balköse, PhD, and
  Sabu Thomas, PhD

- **Chemical Technology and Informatics in Chemistry with Applications**
  Editors: Alexander V. Vakhrushev, DSc, Omari V. Mukbaniani, DSc, and Heru Susanto, PhD

- **Engineering Technologies for Renewable and Recyclable Materials: Physical-Chemical Properties and Functional Aspects**
  Editors: Jithin Joy, Maciej Jaroszewski, PhD, Praveen K. M., and Sabu Thomas, PhD, and Reza Haghi, PhD

- **Engineering Technology and Industrial Chemistry with Applications**
  Editors: Reza Haghi, PhD, and Francisco Torrens, PhD

- **High-Performance Materials and Engineered Chemistry**
  Editors: Francisco Torrens, PhD, Devrim Balköse, PhD, and Sabu Thomas, PhD

- **Methodologies and Applications for Analytical and Physical Chemistry**
  Editors: A. K. Haghi, PhD, Sabu Thomas, PhD, Sukanchan Palit, and Priyanka Main

- **Modern Physical Chemistry: Engineering Models, Materials, and Methods with Applications**
  Editors: Reza Haghi, PhD, Emili Besalú, PhD, Maciej Jaroszewski, PhD, Sabu Thomas, PhD, and Praveen K. M.

- **Physical Chemistry for Chemists and Chemical Engineers: Multidisciplinary Research Perspectives**
  Editors: Alexander V. Vakhrushev, DSc, Reza Haghi, PhD, and J. V. de Julián-Ortiz, PhD

- **Physical Chemistry for Engineering and Applied Sciences: Theoretical and Methodological Implication**
  Editors: A. K. Haghi, PhD, Cristóbal Noé Aguilar, PhD, Sabu Thomas, PhD, and Praveen K. M.

- **Research Methodologies and Practical Applications of Chemistry**
  Editors: Lionello Pogliani, PhD, A. K. Haghi, PhD, and Nazmul Islam, PhD

- **Theoretical Models and Experimental Approaches in Physical Chemistry: Research Methodology and Practical Methods**
  Editors: A. K. Haghi, PhD, Sabu Thomas, PhD, Praveen K. M., and Avinash R. Pai

CONTENTS

CONTRIBUTORS

A. V. Aliev
Federal State Budgetary Educational Institution of Higher Education, Kalashnikov Izhevsk State Technical University, Izhevsk Studencheskaya St. 7, Russia, 426069

Sarmistha Basu
Department of Electronics, Behala College, Kolkata 700060, West Bengal, India

Peter Duchovič
VIPO, Gen. Svobodu 1069/4, SK 958 01 *Partizánske,* Slovakia

Lei Guo
School of Material and Chemical Engineering, Tongren University, Tongren 554300, P.R. China

A. K. Haghi
University of Guilan, Faculty of Engineering, Rasht, Iran

Nazmul Islam
Theoretical and Computational Research Laboratory, Ramgarh Engineering College, Jharkhand 825101, India. E-mail: nazmul.islam786@gmail.com

Ajith James Jose
Department of Chemistry, St. Berchmans College Changanassery, Changanassery 686101, Kerala, India. E-mail: ajithjamesjose@gmail.com

T. K. Jumadilov
JSC "Institute of Chemical Sciences after A.B. Bekturov," Almaty, the Republic of Kazakhstan. E-mail: jumadilov@mail.ru

Peter Jurkovič
VIPO, Gen. Svobodu 1069/4, SK 958 01 *Partizánske,* Slovakia

Savaş Kaya
Department of Chemistry, Faculty of Science, Cumhuriyet University, 58140 Sivas, Turkey

Amjad Mumtaz Khan
Department of Chemistry, Analytical Research Lab, Aligarh Muslim University, Aligarh 202002, Uttar Pradesh, India. E-mail: amjad.mt.khan@gmail.com

Sonia Khanna
Department of Chemistry, School of Basic Science and Research, Sharda University, Greater Noida, India. E-mail: sonia.khanna@sharda.ac.in

Angela Kleinová
Polymer Institute of the Slovak Academy of Sciences, SK 84541 Bratislava, Slovakia

R. G. Kondaurov
JSC "Institute of Chemical Sciences after A.B. Bekturov," Almaty, the Republic of Kazakhstan

P. J. Kurian
Department of Physics, St. Berchmans College, Changanassery 686101, Kerala, India

Yahiya Kadaf Manea
Department of Chemistry, Analytical Research Lab, Aligarh Muslim University, Aligarh 202002, Uttar Pradesh, India

Prijil Mathew
Department of Physics, St. Berchmans College, Changanassery 686101, Kerala, India

Sajith Mathews T.
Department of Physics, St. Berchmans College, Changanassery 686101, Kerala, India

Ján Matyašovský
VIPO, Gen. Svobodu 1069/4, SK 958 01 *Partizánske,* Slovakia

Roman F. Nalewajski
Department of Theoretical Chemistry, Jagiellonian University, Gronostajowa 2, 30–387 Cracow, Poland. E-mail: nalewajs@chemia.uj.edu.pl

Igor Novák
Polymer Institute of the Slovak Academy of Sciences, SK 84541 Bratislava, Slovakia

Lionello Pogliani
Departamento de Química Física, Facultad de Farmacia, Universitat de Valencia, Av. V.A. Estellés s/n, 46100 Burjassot, València, Spain. E-mail: liopo@uv.es

Meha J. Prajapati
Department of Chemistry, Sardar Patel University, Vallabh Vidyanagar 388120, Gujarat, India

Zaki S. Safi
Department of Chemistry, Faculty of Science, Al Azhar University-Gaza, Gaza City, P.O. Box 1277, Palestine

A. S. M. Saleh
College of Grain Science and Technology, Shenyang Normal University, Shenyang, Liaoning 110034, China
Department of Food Science and Technology, Faculty of Agriculture, Assiut University, Assiut, Egypt

Ravindra S. Shnide
Department of Chemistry and Industrial Chemistry, Dayanand Science College, Latur 413512, Maharashtra, India

Ladislav Šoltes
Centre of Experimental Medicine, Institute of Experimental Pharmacology and Toxicology, Slovak Academy of Sciences, SK 84104 Bratislava, Slovakia

Kiran R. Surati
Department of Chemistry, Sardar Patel University, Vallabh Vidyanagar 388120, Gujarat, India. E-mail: kiransurati@yahoo.co.in

Heru Susanto
Computational Science, The Indonesian Institute of Sciences, Indonesia
Computer Science Department, Tunghai University, Taiwan. E-mail: susanto.net.id@gmail.com

T. M. Tamer
College of Grain Science and Technology, Shenyang Normal University, Shenyang,
Liaoning 110034, China
Polymer Materials Research Department, Advanced Technologies and New Materials Research
Institute (ATNMRI), City of Scientific Research and Technological Applications (SRTA-City),
New Borg El-Arab City, 21934, Alexandria, Egypt

Katarína Valachová
Centre of Experimental Medicine, Institute of Experimental Pharmacology and Toxicology,
Slovak Academy of Sciences, SK 84104 Bratislava, Slovakia

O. A. Voevodina
Federal State Budgetary Educational Institution of Higher Education, Kalashnikov Izhevsk State
Technical University, Izhevsk Studencheskaya St. 7, Russia, 426069

N. Wang
College of Grain Science and Technology, Shenyang Normal University, Shenyang,
Liaoning 110034, China

P. Wang
College of Grain Science and Technology, Shenyang Normal University, Shenyang,
Liaoning 110034, China

Z. G. Xiao
College of Grain Science and Technology, Shenyang Normal University, Shenyang,
Liaoning 110034, China

Ebru Yabaş
Department of Chemistry and Chemical Processing Technologies,
Cumhuriyet University Imranlı Vocational School, 58980 Sivas, Turkey

L. Yang
College of Grain Science and Technology, Shenyang Normal University, Shenyang,
Liaoning 110034, China

Q. Y. Yang
College of Grain Science and Technology, Shenyang Normal University, Shenyang,
Liaoning 110034, China

ABBREVIATIONS

ABTS	2, 2'-azino-bis (3-ethylbenzothiazoline-6-sulphonic acid
ADMET	absorption, distribution, metabolism, excretion, and toxicity
AHCA	agglomerative hierarchical cluster analysis
AIM	atoms-in-molecules
CA	cellulose acetate
CADD	computer analysis drug design
CASD	computer-assisted synthesis design
CDFT	conceptual density functional theory
COMPASS	condensed phase optimized molecular potentials for atomistic simulation studies
DFT	density functional theory
DPPH	2,2-diphenyl-1-picrylhydrazyl
EPM	electrostatic potential map
ESP	electrostatic potential map
FMO	frontier molecular orbital
FTIR	Fourier-transform infrared
FTIR-ATR	Fourier-transform infrared spectroscopy with attenuated total reflectance
GD-AAS	glow discharges for atomic absorption spectrometry
GD-OES	glow discharge optical emission spectrometry
GDMS	glow discharge mass spectrometry
HOMO	highest occupied molecular orbital
hP2M5VP	poly-2-methyl-5-vinylpyridine hydrogel
hPMAA	polymethacrylic acid hydrogel
HTS	high throughput screening
ITER	International Thermonuclear Experimental Reactor
LB	Luria-Bertani medium
LMICs	low- and middle-income countries
LUMO	lowest unoccupied molecular orbital
MACCS	Molecular ACCess System
MLCT	metal-to-ligand charge transfer
MMDB	molecular modelling databases

NHC	N-heterocyclic carbenes
OLEDs	organic light-emitting diodes
P2M5VP	poly-2-methyl-5-vinylpyridine
PA	proton affinity
PCA	principal component analysis
PFC	phloroglucinol–formaldehyde composite
PMAA	polymethacrylic acid
QM	quantum mechanics
QSAR	quality structure activity relationship
RFD	radio-frequency discharge
RS	reactive species
SAR	structure–activity relationship
SE	Schrödinger equation
SEE	surface energy evaluation
SMILES	simplified molecular input line entry system
TS	tensile strength
WOLEDs	white organic light emitting diodes
XRD	X-ray diffraction

PREFACE

This book presents a detailed analysis of current experimental and theoretical approaches surrounding chemical science. The research in this book will provide experimentalists, professionals, students, and academicians with an in-depth understanding of chemistry and its impact on modern technology. It also provides a comprehensive overview of theoretical and experimental chemistry while focusing on the basic principles that unite the subdisciplines of the field. With an emphasis on multidisciplinary, as well as interdisciplinary applications, the book extensively reviews fundamental principles and presents recent research to help the reader make logical connections between the theory and application of modern chemistry concepts. It also emphasizes the behavior of material from the molecular point of view for postgraduate students who have a background in chemistry and physics and in thermodynamics.

The burgeoning field of chemistry and chemical science has led to many recent technological innovations and discoveries. Understanding the impact of these technologies on business, science, and industry is an important first step in developing applications for a variety of settings and contexts.

The aim of this book is to present research that has transformed this discipline and aided its advancement. The readers would gain knowledge that would broaden their perspective about the subject.

The book examines the strengths and future potential of chemical technologies in a variety of industries. Highlighting the benefits, shortcomings, and emerging perspectives in the application of chemical technologies, this book is a comprehensive reference source for chemists, engineers, graduate students, and researchers with an interest in the multidisciplinary applications, as well as the ongoing research in the field.

PREFACE

This book presents a detailed analysis of current experimental and theoretical approaches surrounding chemical science. The research in this book will provide experimentalists, professionals, students, and academicians with an in-depth understanding of chemistry and its impact on modern technology. It also provides a comprehensive overview of theoretical and experimental chemistry while focusing on the basic principles that unite the subdisciplines of the field. With an emphasis on multidisciplinary, as well as interdisciplinary applications, the book extensively reviews fundamental principles and presents recent research to help the reader make logical connections between the theory and application of modern chemistry concepts. It also emphasizes the behavior of material from the molecular point of view for postgraduate students who have a background in chemistry and physics and in thermodynamics.

The burgeoning field of chemistry and chemical science has led to many recent technological innovations and discoveries. Understanding the impact of these technologies on business, science, and industry is an important first step in developing applications for a variety of settings and contexts.

The aim of this book is to present research that has transformed this discipline and aided its advancement. The readers would gain knowledge that would broaden their perspective about the subject.

The book examines the strengths and future potential of chemical technologies in a variety of industries. Highlighting the benefits, shortcomings, and emerging perspectives in the application of chemical technologies, this book is a comprehensive reference source for chemists, engineers, graduate students, and researchers with an interest in the multidisciplinary applications, as well as the ongoing research in the field.

CHAPTER 1

SOME REMARKS ABOUT PSEUDO-ZERO-ORDER REACTIONS

LIONELLO POGLIANI*

Departamento de Química Física, Facultad de Farmacia, Universitat de Valencia, Av. V.A. Estellés s/n, 46100 Burjassot, València, Spain

E-mail: liopo@uv.es

ABSTRACT

Experimental evidence has shown that reactions at the very first instant of their inception, that is, short after their start, follow a pseudo-zero-order kinetics. The initial rate method that studies only the first instants of a reaction allows us to analyze any elementary step of a chemical reaction with a pseudo-zero-order kinetics. To derive valuable information about the real kinetic properties of a reaction, it should be run and studied at different initial concentrations. The economy in time required to collect data for a pseudo-zero reaction is consistent.

1.1 INTRODUCTION

In 2008, Pogliani[1] explained that any elementary step of a chemical reaction in the initial rate approximation (i.e., short after its inception) can, with a good approximation, be analyzed as a pseudo-zero-order kinetics. In fact, at this condition, any higher-order kinetics[2] cannot be differentiated from a zero-order kinetics. Quite recently, experimental evidence was found for the detection of pseudo-zero-order kinetics under the conditions suggested by Pogliani.[1] The first experimental study published in 2016[3] stated that the pseudo-zero-order reaction kinetics fitted quite well with the very first moments of a reaction. In this study, it is reported that the kinetic data for elementary mercury absorption collected over a short period, followed by a zero-order or pseudo-zero-order kinetics. The second experimental study[4] found that kinetic data for the neutralization of 2,2-diphenyl-1-picrylhydrazyl (DPPH) in the cathodic compartment of microbial fuel cell dual chambers followed a pseudo-zero-order kinetics. This fact was not the result of a particular mechanism of the reaction, but an artifact of the conditions under which the reaction was carried out, as showed by Pogliani.[1] A third and quite recent study[5] found out that the kinetics of mass loss of 2,4,6-trinitrotoluene (TNT) explosive samples in isothermal conditions could normally be calculated with the kind of pseudo-zero-order kinetics outlined by Pogliani.[1] These last authors go on, after the calculation of the pseudo-zero-order kinetic constants at different temperatures, to derive with the Arrhenius equation the activation energy of the mass loss process.

Due to the increasing importance of the subject in the following paragraphs, we will review and expand the original paper about this kind of reaction kinetics.

1.2   GENERAL METHODOLOGICAL CONSIDERATIONS

Normally, the analysis of kinetic data can be quite challenging, if the number of data is limited to no more than two, and in some cases even three half-lives. Reactions examined for periods shorter than two half-lives hardly allow us to define their real kinetic characteristics and the use of the integrated rate laws is of no help. Zero-order reactions are usually underrated in chemical kinetic studies and are a secondary subject in physical chemistry and chemical kinetics textbooks. The fact that pseudo-zero order, like the zero-order kinetics do not require the elaborated mathematics that higher-order reactions do,[2] could be used to analyze kinetic data collected over short periods when the concentration of reactants has hardly changed. It should be remarked that the initial rate method, a commonly used method in chemical kinetics, when applied to the rate laws, allows to reduce the order of a higher-order reaction kinetics from the second or higher order into a pseudo-first order.[2]

It will be shown here that in the limit of very short times (or at high reactant concentrations), any elementary reaction step mimics a pseudo-zero-order kinetics. Clearly, depending on the value of the kinetic constant k, "short times" could actually mean a fraction of a second, seconds, minutes, hours, or even days or years (i.e., with increasingly small k). In this limit, the integrated rate laws simplify into pseudo-zero-order rate laws whose real characteristics can be uncovered by collecting data at different initial concentrations.

1.3   MATHEMATICAL METHOD

A zero-order reaction has the following rate, and integrated rate laws,[2]

$$-d[A]/dt = k \ \text{ and } \ [A] = [A_0] - kt \tag{1.1}$$

$[A_0]$ is the initial reactant concentration. A first-order elementary step reaction rate has the following rate, and integrated rate laws[2]

$$-d[A]/dt = k[A] \ \text{ and } \ [A] = [A_0]\exp(-kt) \tag{1.2}$$

For $kt \ll 1$ (either k or t or both too small), this equation (with the McLaurin series expansion for the exponential, and neglecting all but the first two terms) can be transformed into the following pseudo-zero-order integrated rate law where, $k' = k[A_0]$,

$$[A] = [A_0](1 - kt) = [A_0] - k't \tag{1.3}$$

Notice that either for $[A_0]$ too big or kt too small $[A] \cong [A_0]$, and $-d[A]/dt = k' = k[A_0]$. The second order rate and integrated rate laws are,[2]

$$-d[A]/dt = k[A]^2 \ \text{ and } \ [A] = [A_0]/(1 + kt[A_0]) \tag{1.4}$$

With the simplified McLaurin series expansion for kt quite small, such that $kt\,[A_0] \ll 1$, we obtain the following pseudo-zero-order rate law, where $k'' = k[A_0]^2$ (notice that the rate law in this limit becomes: $-d[A]/dt = k'' = k[A_0]^2$).

$$[A] = [A_0](1 - kt[A_0]) = [A_0] - k''t \tag{1.5}$$

Similarly, for the other type of second-order kinetics,[2]

$$-d[A]/dt = k[A][B] \quad \text{and} \quad [A]/[B] = \exp\{-kt([B_0] - [A_0]\} \cdot [A_0]/[B_0]$$
$$(1.6)$$

The condition kt quite small, such that $kt([B_0] - [A_0]) < 1$, gives rise to,

$$[A]/[B] = [1 - k'''t] \cdot [A_0]/[B_0] \quad \text{with } k''' = k([B_0] - [A_0]) \qquad (1.7)$$

With $[A]/[B] = Y$, $C = [A_0]/[B_0]$, and $Ck''' = k^{iv}$, this equation becomes the well-known integrated rate law for a pseudo-zero-order kinetics,

$$Y = C (1 - k'''t) = C - k^{iv}t \qquad (1.8)$$

Elementary reaction steps with order three or higher are quite unlikely to occur. Anyway, the general case for n^{th}-order reactions shows the following rate and integrated rate laws (notice that $n > 1$),[2]

$$-d[A]/dt = k[A]^n \quad \text{and} \quad [A]^{n-1} = [A_0]^{n-1}/\{1 + kt(n-1)[A_0]^{n-1}\} \qquad (1.9)$$

For $kt \ll 1$, such that $kt(n-1)[A_0]^{n-1} \ll 1$, we have the following pseudo-zero-order kinetics, where: $^nk = k(n-1)[A_0]^{(n-1)}$,

$$[A]^{n-1} = [A_0]^{n-1}\{1 - kt(n-1)[A_0]^{(n-1)}\} = [A_0]^{n-1} - {}^nkt \qquad (1.10)$$

1.4   KINETIC PLOTS

Figure 1.1 shows in a dimensionless unit ($t^0 \equiv 1$ in any unit, thus t/t^0 is dimensionless) the simulated data for the first instants of a zero-order ($\bullet$), first-order ($\blacklozenge$), and second-order ($\blacktriangle$) reactions. The three sets of kinetic data were generated with a $k/k^0 = 0.01$ rate constant ($k^0 \equiv 1$ in any unit, thus k/k^0 is dimensionless), and $[A_0] = 1.5$ mol L^{-1}. The data were computed using the original integrated rate laws (eqs 1.1, 1.2, and 1.4).

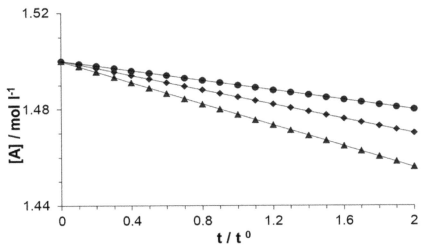

FIGURE 1.1 Kinetic data obtained with a k/k^0 value of 0.01, with $[^1A_0] = 1.5$ mol L^{-1} for zero-order (●), first-order (♦) and second-order (▲) kinetics.

Let us now consider that the data in Figure 1.1 are the result of three different experimental kinetic measurements due to three different reactions. Least squares analysis for the three kinetic data tells that we face with a great precision (r^2 = regression correlation coefficient) three linear relationships, a characteristic shared either by the zero or pseudo-zero-order kinetics. From these linear relations, it is possible to extract the value for the three slopes, that is, for the three rate constants, k, k', and k'',

●) $[A] = -0.01 \cdot t + 1.5$; $r^2 = 1$, $k = 0.01$ mol·L^{-1}s^{-1}
♦) $[A] = -0.0149 \cdot t + 1.5$; $r^2 = 1$, $^1k' = 0.0149$ mol·L^{-1}s^{-1}
▲) $[A] = -0.0218 \cdot t + 1.4999$; $r^2 = 0.9999$, $^1k'' = 0.0218$ mol·L^{-1}s^{-1}

Actually, we (not the experimenter who is collecting the data) know that: $^1k' = k[^1A_0]$, and $^1k'' = k[^1A_0]^2$, whereas k is always independent of $[A_0]$ and it is the true rate constant of a reaction kinetics. All present data, even if two of them were obtained with nonlinear relations, obey a linear zero or pseudo-zero-order kinetics and no consistent deviations can be detected that could tell us that we are facing two nonlinear kinetics. Any kinetic analysis of one of the three cases could not uncover the real kinetic character of the data. An investigator who is studying the kinetics of a chemical reaction in the initial rate approximation can only conclude that the reaction obeys a pseudo-zero-order kinetics with rate constant, k_{exp}. To solve the question about the true character of a pseudo-zero elementary reaction, that is, if it is either a zero, a first, or a second (or even higher) order reaction and if the true value for k_{exp} is either k, $^1k'$, or $^1k''$, at the initial rate approximation, the reaction should be run, at least, at a different $[^2A_0]$ value.

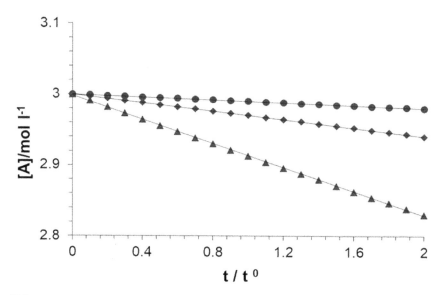

FIGURE 1.2 Zero-order (●), first-order (♦), and second-order (▲) kinetics, obtained with $[^2A_0] = 3.0$ mol L^{-1} with a $k/k^0 = 0.01$ (like the previous data).

Figure 1.2 shows the same types of reaction kinetics, this time with $[A_0] = 3$ mol L^{-1}. Least squares analysis shows that, with a good precision, a linear relationship holds for the three kinetic data, from which we can derive the values for the three rate constants, k, $^2k'$, and $^2k''$,

●) $[A] = -0.01 \cdot t + 3.000$; $r^2 = 1$, that is, $k = 0.01$ mol·L^{-1}s^{-1}
♦) $[A] = -0.0297 \cdot t + 2.9999$; $r^2 = 1$, that is, $^2k' = 0.0297$ mol·L^{-1}s^{-1}
▲) $[A] = -0.0849 \cdot t + 2.9984$; $r^2 = 0.9998$, that is, $^2k'' = 0.0849$ mol·L^{-1}s^{-1}

We are again facing three pseudo-zero-order kinetics with slopes, k, $^2k'$, and $^2k''$, where, $^2k' = k[^2A_0]$ and $^2k'' = k[^2A_0]^2$. Now, contrasting the results from the two figures an investigator, who is studying a reaction kinetics, can either discard or confirm that k_{exp} for the two ($\bullet$) curves do not depend on $[A_0]$ and eventually conclude that the pseudo-zero-order kinetics is of the zero order type with $^1k_{exp} = \,^2k_{exp} = k$. If the reaction data do not fit this case, then the experimenter goes over to the next two cases. If the experimental data fit the ($\blacklozenge$)-type relations of Figures 1.1 and 1.2, then $^1k_{exp} = k[^1A_0] (= \,^1k')$, and $^2k_{exp} = k[^2A_0] (= \,^2k')$ hold, and the pseudo-zero reaction encodes a first-order kinetics with rate constant k. If, instead, the data fit the ($\blacktriangle$)-type relations of Figures 1.1 and 1.2, then $^1k_{exp} = k[^1A_0]^2 (= \,^1k'')$, and $^2k_{exp} = k[^2A_0]^2 (= \,^2k'')$ hold, and the pseudo-zero elementary step encodes a second-order kinetics with rate constant k. Thus, with this method, it is rather easy to retrieve the true k and the real order at $kt \ll 1$.

Notice that the slope increases with the order of a reaction for $[A_0] > 1$ (see eq 1.10, where $^nk = k(n-1)[A_0]^{(n-1)}$ and eqs 1.1, 1.3, and 1.5), whereas for $[A_0] < 1$, excluding the zero-order kinetics, the slopes decrease and get inverted. Considering eq 1.10 for $n = 3$, when $2kt[A_0]^2 < 0.1$ no deviation from a linear relationship could be detected in a $[A]$ vs. t plot. The higher the order, the more the condition $kt(n-1)[A_0]^{n-1} < 0.1$ should hold if a pseudo-zero-order behavior is to be detected.

1.5 CONCLUSION

Chemical kinetics of elementary reaction steps in the initial rate approximation (for $kt \ll 1$, when $[A] \approx [A_0]$) always follow a pseudo-zero-order kinetics. Nevertheless, studying reactions nearly at their "birth-time" at different initial concentrations, it is possible to identify the true character of a pseudo-zero-order reaction. It should be underlined that the study of elementary steps at the initial rate approximation simplifies the mathematical formalism needed to analyze the data, and helps to shorten the time needed to collect them. The changes that integrated rate laws undergo for small kt is an aspect of a well-known and more general problem in science. For instance, relativistic mechanics, practically, goes over into Newtonian mechanics for speeds far from the speed of light (and far from hugely massive bodies), while for macroscopic bodies classical mechanics is more effective than quantum mechanics.

KEYWORDS

- experimental results
- initial rate approximation
- rate laws and integrated rate laws
- pseudo-zero-order kinetics
- concentration

REFERENCES

1. Pogliani, L. Pseudo-zero-order Reactions. *React. Kinet. Catal. Lett.* **2008,** *93,* 187–191.
2. Atkins, P. W.; De Paula, J. *Physical Chemistry.* Oxford University Press: Oxford, 2002.
3. Khairiraihanna, J.; Norasikin, S.; Shiow, T. S.; Siew, C. C.; Helen, K.; Hanapi, M. Development of Coconut Pith Chars Towards High Elemental Mercury Adsorption Performance—Effect of Pyrolysis Temperatures. *Chemosphere* **2016,** *156,* 56–68.
4. Ralitza, K.; Hyusein Y.; Valentin N. Microbial Fuel Cell as a Free-radical Scavenging Tool. *Biotechnol. Biotechnol. Equipment* **2017,** *31,* 511–515.
5. Hamid, R. P.; Sajjad, D.; Parvaneh, N.; Ehsan, F. G. The Kinetic of Mass Loss of Grades A and B of Melted TNT by Isothermal and Non-isothermal Gravimetric Methods. *Defence Technol.* **2018,** *14,* 126–131.

CHAPTER 2

ON CLASSICAL AND QUANTUM ENTROPY/INFORMATION DESCRIPTORS OF MOLECULAR ELECTRONIC STATES[1]

ROMAN F. NALEWAJSKI*

Department of Theoretical Chemistry, Jagiellonian University, Gronostajowa 2, 30–387 Cracow, Poland

E-mail: nalewajs@chemia.uj.edu.pl

ABSTRACT

The modulus (probability) and phase (current) components of complex quantum states generate the resultant descriptors of the information/entropy content in molecular wavefunctions. These overall informa-tion–theoretic concepts combine the classical (probability) contributions of the gradient information or global entropy, and the corresponding nonclassical (phase/current) supplements. The densities of classical and nonclassical components of the generalized Fisher (information) and Shannon (entropy) measures obey identical mutual relations. These resul-tant measures of the information and entropy content in the pure quantum state represent expectations of the Hermitian and non-Hermitian operators, respectively, giving rise to real and complex average values. The complex entropy concept combines the classical and nonclassical components as

---

[1]The following notation is adopted throughout: A denotes a *scalar*, A is the row/column *vector*, **A** represents a square or rectangular *matrix*, and the dashed symbol $\hat{A}$ stands for the quantum-mechanical *operator* of the physical property A. The logarithm of the Shannon information measure is taken to an arbitrary but fixed base: $\log = \log_2$ corresponds to the information content measured in *bits* (binary digits), while $\log = \ln$ expresses the amount of information in *nats* (natural units): 1 nat = 1.44 bits.

its real and imaginary parts. A relation between this novel measure and von Neumann's quantum entropy in density matrix is examined. Both measures are shown to conform to general analogies between the classical and quantum descriptions of molecular states. The (complex) resultant entropy explicitly accounts for the information content in the phase/current component of molecular states; this contribution is cancelled out in the (real) measure of von Neumann.

2.1   INTRODUCTION

The information-theoretic (IT) approach[1–5] has proven its utility in a variety of molecular applications, for example, Refs. [6–16]. In molecular quantum mechanics (QM), a general electronic wavefunction is a complex entity characterized by its modulus and phase components. The square of the former defines the particle probability distribution, the system static structure of "being," while the gradient of the latter generates the current density, the state dynamical structure of "becoming."[17] The continuity of probability density relates these two structural aspects: the probability dynamics is determined by current's divergence. These two structural facets generate the associated classical (probability) and nonclassical (phase/current) contributions to the resultant measures of the information/entropy content in complex electronic states.[13,18–21] The densities of these overall entropy/information measures satisfy classical relations characterizing the classical functionals of Fisher[1] and Shannon.[2] For example, the complex entropy[13,21] combines the classical (real) and nonclassical (imaginary) contributions due to the state probability (wavefunction modulus) and current (wavefunction phase), respectively. Such generalized concepts allow one to distinguish the information content of states generating the same electron density but differing in their phase/current composition. They also allow a more precise phase description of the bonding status of molecular fragments,[22–24] for example, the entangled states[25] of subsystems.[26] The entropic principles using such resultant IT descriptors of electronic states have been also used to determine the information equilibria in molecular systems and their constituent parts.[13,27–32] The phase aspect of molecular states is also vital for the quantum (amplitude) communications between atoms in molecules,[9,10,12,13,15] which determine entropic descriptors of the chemical bond multiplicities and their covalent/ionic composition.[13,33–36] The noadditive Fisher information[1,8,37] has been shown to be crucial for localizing electrons and bonds.[37,38]

The present overview first examines the modulus (probability) and phase (current) degrees-of-freedom of general (complex) quantum states, summarizes their continuity relations, and introduces the resultant entropy/information descriptors contained in the pure quantum state. The quantum operators of the classical (probability) and nonclassical (phase/current) contributions will be introduced and the mutual relations between components of the densities-per-electron of the gradient (determinacy) information and global (indeterminacy) entropy will be elucidated. A relation of the novel complex entropy to the quantum von Neumann's entropy[39] will be examined and general similarities between the classical and quantum descriptions of the time variations of the mixed-state density operators will be elucidated in quantum ensembles. It will be demonstrated that the novel (complex) entropy concept conforms to these general analogies. In the (real) approach of von Neumann, the phase contribution to the overall information content in complex molecular states is shown to be exactly cancelled out.

For simplicity reasons, the one-electron case is assumed throughout. However, the modulus (density) and the phase (current) aspects of general electronic states can be similarly separated using the Harriman–Zumbach–Maschke construction[40,41] of Slater determinants yielding the specified electron density. In this treatment, the common modulus part of N occupied (orthonormal) Equidensity Orbitals, for the given ground-state density ρ, reflects the molecular probability distribution $p(r) = \rho(r)/N$, and so do the density optimum (orthonormal) orbitals in the Kohn–Sham[42] implementation of density functional theory,[43] describing the associated separable (noninteracting) system. In fact the two sets constitute equivalent sets of spin-orbitals linked by the unitary transformation and generating identical Slater determinants and classical entropy/information measures.

2.2   PROBABILITY AND CURRENT COMPONENTS OF MOLECULAR STATES

Let us consider a single electron system ($N = 1$) in quantum state $|\psi(t)\rangle \equiv |\psi(t)\rangle$ generating the complex wave function

$$\psi(\mathbf{r}, t) = \langle \mathbf{r}|\psi(t)\rangle = R(\mathbf{r}, t) \exp[i\phi(\mathbf{r}, t)], \tag{2.1}$$

with $R(\mathbf{r}, t)$ and $\phi(\mathbf{r}, t)$ standing for its modulus and phase components. It determines the particle probability distribution $p(\mathbf{r}, t) = \psi(\mathbf{r}, t)^* \psi(\mathbf{r}, t) = R(\mathbf{r}, t)^2$ and the current density

$$\mathbf{j}(\mathbf{r}, t) = (\hbar/2mi)[\psi(\mathbf{r}, t)^* \nabla \psi(\mathbf{r}, t) - \psi(\mathbf{r}, t) \nabla \psi(\mathbf{r}, t)^*]$$

$$= (\hbar/m)] \, p(\mathbf{r}, t) \nabla \phi(\mathbf{r}, t) \equiv p(\mathbf{r}, t) V(\mathbf{r}, t), \tag{2.2}$$

where the velocity field of probability "fluid" reflects the state phase-gradient $\nabla \phi(\mathbf{r}, t)$ measuring the current-per-particle: $V(\mathbf{r}, t) = \mathbf{j}(\mathbf{r}, t)/p(\mathbf{r}, t)$. In molecular scenario, the electron moves in the external potential $v(\mathbf{r})$ due to nuclei, in the electronic Hamiltonian

$$\hat{H}(\mathbf{r}) = -(\hbar^2/2m)\nabla^2 + v(\mathbf{r}) \equiv \hat{T}(\mathbf{r}) + v(\mathbf{r}). \tag{2.3}$$

The state dynamics is determined by Schrödinger equation (SE)

$$i\hbar \, \partial \psi(\mathbf{r}, t)/\partial t = \hat{H}(\mathbf{r}) \, \psi(\mathbf{r}, t) \text{ or} \tag{2.4}$$

$$\psi(\mathbf{r}, t) = \exp[-i\hbar^{-1} \hat{H}(\mathbf{r}) \, t] \psi(\mathbf{r}, t = 0). \tag{2.5}$$

It further implies specific temporal evolutions of the instantaneous probability density $p(\mathbf{r}, t)$ and the state phase $\phi(\mathbf{r}, t)$. The time derivative of the former measures the vanishing probability-source σp in the continuity relation

$$\sigma_p(\mathbf{r}, t) \equiv dp(\mathbf{r}, t)/dt = \partial p(\mathbf{r}, t)/\partial t + \nabla \times \mathbf{j}(\mathbf{r}, t) = 0 \quad \text{or} \tag{2.6}$$

$$\partial p(\mathbf{r}, t)/\partial t = -\nabla \times \mathbf{j}(\mathbf{r}, t). \tag{2.7}$$

In the statistical mixture of quantum states $\{\psi_\alpha\}$ observed with (external) probabilities $\{p_\alpha\}$, $\sum_\alpha p_\alpha = 1$, defined by the density operator

$$\hat\rho(t) = \sum_\alpha |\psi_\alpha(t)\rangle p_\alpha(t) \langle \psi_\alpha(t)|, \qquad (2.8)$$

the ensemble probabilities are preserved in time, $p_\alpha(t) = p_\alpha(t=0)$ and the dynamics of $\hat\rho$ is determined by SE:

$$i\hbar \partial \hat\rho(t)/\partial t = [\hat H, \hat\rho(t)] \equiv \hat L \, \hat\rho(t). \qquad (2.9)$$

Here, $\hat L$ stands for Liouville's operator of Prigigine[17] in terms of which the formal solution of the preceding equation reads:

$$\hat\rho(t) = \exp[-i\hbar^{-1} \hat L t] \, \hat\rho(t=0). \qquad (2.10)$$

This equation provides the formal quantum analog of the familiar Liouville equation of classical mechanics for the dynamics of the phase-space density $\rho(r, p, t)$ of the Gibbs ensemble,

$$i\,[\partial\rho/\partial t] = i\{H, \rho\} \equiv \hat{\mathcal{L}} \rho, \qquad (2.11)$$

where $H(r, p, t)$ stands for the classical Hamilton's function, the system energy expressed in terms of its spatial ($r = \{q_i\}$) and momentum ($p = \{p_i\}$) degrees of freedom, the Poisson bracket

$$\{H, \rho\} = -\sum_i [(\partial H/\partial p_i)\,(\partial\rho/\partial q_i) - (\partial H/\partial q_i)\,(\partial\rho/\partial p_i)], \qquad (2.12)$$

and the Liouville operator

$$\hat{\mathcal{L}} = -i[(\partial H/\partial p)\times(\partial/\partial r) - (\partial H/\partial r)\times(\partial/\partial p)]. \qquad (2.13)$$

The formal solution of eq 2.11 thus reads [compare eq 2.10]:

$$\rho(t) = \exp[-i\,\hat{\mathcal{L}}\,t]\,\rho(0). \qquad (2.14)$$

The velocity descriptor can be also attributed to the current concept associated with the phase component, $J(r, t) = \phi(r, t) V(r, t)$, determining a nonvanishing phase-source in the associated continuity equation:

$$\sigma_\phi(r, t) \equiv d\phi(r, t)/dt = \partial\phi(r, t)/\partial t + \nabla \times J(r, t) \qquad \text{or} \qquad (2.15)$$

$$\partial\phi(r, t)/\partial t - \sigma_\phi(r, t) = -\nabla \times J(r, t). \qquad (2.16)$$

The local phase dynamics from SE,

$$\partial/\partial t = [\hbar/(2m)] [R^{-1}\Delta R - (\nabla\phi)^2] - v/\hbar, \qquad (2.17)$$

then identifies the phase source:

$$\sigma_\phi = [\hbar/(2m)] [R^{-1}\Delta R + (\nabla)^2] - v/\hbar. \qquad (2.18)$$

For example, in the eigenstate of the Hamiltonian corresponding to energy E_s, $\psi_s(r, t) = R_s(r)\exp[i\phi_s(t)]$, where $\phi_s(r, t) = \phi_s(t) = -(E_s/\hbar)t \equiv -\omega_s t$ and hence $j_s(r, t) = V_s(r, t) = J_s(r, t) = 0$, the stationary SE for the probability amplitude $R_s(r, t) = R_s(r)$, $p_s(r, t) = R_s(r)^2 = p_s(r)$,

$$\hat{H}(r) R_s(r) = -(\hbar^2/2m) \Delta R_s(r) + v(r) R_s(r) = E_s R_s(r), \qquad (2.19)$$

and the phase dynamics of eq 2.17 identify the constant phase source:

$$\sigma_\phi = (\partial\phi_s/\partial t) = [\hbar/(2m)] (R_s^{-1}\Delta R_s) - v = -\omega_s. \qquad (2.20)$$

2.3 CLASSICAL AND RESULTANT ENTROPY/INFORMATION CONCEPTS

At instant $t = t_0 = 0$ the average Fisher (F)[1] measure of the classical gradient information for locality events contained in probability density $p(r) = R(r)^2$ is reminiscent of von Weizsäcker's[44] inhomogeneity correction to the kinetic-energy functional in Thomas–Fermi theory,

$$I[p] = \int [\nabla p(r)]^2/p(r)\, dr = \int p(r) [\nabla \ln p(r)]^2\, dr \equiv I^F[p]$$

$$\equiv \int p(\mathbf{r}) I_p(\mathbf{r}) \, d\mathbf{r} \equiv \int \mathcal{I}_p(\mathbf{r}) \, d\mathbf{r} = 4\int [\nabla R(\mathbf{r})]^2 \, d\mathbf{r} \equiv I[R]. \tag{2.21}$$

Here, $\mathcal{I}_p(\mathbf{r}) = p(\mathbf{r}) I_p(\mathbf{r})$ denotes the functional density and $I_p(\mathbf{r})$ stands for the associated density-per-electron. The amplitude form $I[R]$ reveals that this classical descriptor of the position determinacy measures the average length of the modulus gradient ∇R. Accordingly, the Shannon $(S)^2$ descriptor of the classical global entropy in $p(\mathbf{r})$,

$$S[p] = -\int p(\mathbf{r}) \ln p(\mathbf{r}) \, d\mathbf{r} \equiv \int p(\mathbf{r}) S_p(\mathbf{r}) \, d\mathbf{r} \equiv \int \mathcal{S}_p(\mathbf{r}) \, d\mathbf{r} = -2\int R^2(\mathbf{r}) \ln R(\mathbf{r}) \, d\mathbf{r} \equiv S[R], \tag{2.22}$$

reflects the distribution "spread" (uncertainty), that is, the position inde-terminacy. It also provides the amount of information received, when this uncertainty is removed by an appropriate particle-localization experiment: $I^S[p] \equiv S[p]$.

The densities-per-electron of these complementary information and entropy functionals are seen to satisfy the classical relation

$$I_p(\mathbf{r}) = [\nabla S_p(\mathbf{r})]^2 \tag{2.23}$$

These classical probability functionals generalize naturally into the corresponding resultant quantum descriptors, which combine the prob-ability and phase/current contributions to the overall entropy/information content in the specified electronic state.[13,18–21] Such concepts are applicable to complex wavefunctions of molecular QM. They are defined as expec-tation values of the associated operators: the Hermitian operator of the gradient information,

$$\hat{I}(\mathbf{r}) = -4\Delta = (2i\nabla)^2 = (8m/\hbar^2) \, \hat{T}(\mathbf{r}), \tag{2.24}$$

related to the kinetic energy operator $\hat{S}(\mathbf{r})$ in eq 2.3, and the non-Hermitian operator of the complex entropy,[21]

$$\hat{S}(\mathbf{r}) = -2\ln \psi(\mathbf{r}) = -[\ln p(\mathbf{r}) + 2i\phi(\mathbf{r})] \equiv \hat{S}_p(\mathbf{r}) + \hat{S}_\phi(\mathbf{r}). \tag{2.25}$$

The (real) overall gradient information combines the classical (prob-ability) and nonclassical (phase/current) contributions:

$$I[\psi] = \langle \psi | \hat{I} | \psi \rangle = 4 \int |\nabla \psi(r)|^2 \, dr \equiv \int p(r) I_\psi(r) \, dr \equiv \int \mathcal{I}_\psi(r) \, dr$$

$$= I[p] + 4 \int p(r) [\nabla \phi(r)]^2 \, dr \equiv I[p] + I[\phi] \equiv I[p, \phi]$$

$$= I[p] + (2m/\hbar)^2 \int p(r)^{-1} j(r)^2 \, dr \equiv I[p] + I[j] \equiv I[p, j]. \qquad (2.26)$$

The (complex) resultant entropy is similarly determined by the following (classical) and imaginary (nonclassical) contributions:

$$M[\psi] = \langle \psi | \hat{T} | \psi \rangle = -2 \int p(r) \ln \psi(r) \, dr \equiv \int p(r) M_\psi(r) \, dr \equiv \int \mathcal{M}_\psi(r) \, dr$$

$$= S[p] - 2i \int p(r) \phi(r) \, dr \equiv S[p] + iS[\phi] \equiv M[p] + M[\phi] \equiv M[p, \phi]. \qquad (2.27)$$

The resultant gradient information $I[\psi]$ is thus proportional to the state average kinetic energy $T[\psi] = \langle \psi | \hat{T} | \psi \rangle = (\hbar^2/8m) \, I[\psi]$. The resultant densities-per-electron of these functionals,

$$I_\psi(r) = I_p(r) + 4[\nabla \phi(r)]^2 \equiv I_p(r) + I_\phi(r)$$

$$= I_p(r) + \{(2m/\hbar) j(r)/p(r)\}^2 = I_p(r) + [(2m/\hbar) V(r)]^2 \equiv I_p(r) + I_j(r), \qquad (2.28)$$

$$M_\psi(r) = S_p(r) - 2i\phi(r) \equiv S_p(r) + iS_\phi(r), \qquad (2.29)$$

now satisfy the complex-generalized relation of eq 2.13

$$I_\psi(r) = |\nabla H_\psi(r)|^2 = [\nabla S_p(r)]^2 + [\nabla S_\phi(r)]^2. \qquad (2.30)$$

The complex entropy $M[\psi]$ of the resultant global entropy in quantum state thus provides a natural "vector" generalization of the "scalar" Shannon measure $S[p]$ of the global entropy content in probability distribution. This expectation value of the *non*-Hermitian operator of eq 2.25 generates the probability and phase components of $M[\psi]$ as its real and imaginary parts.

The global entropy of the probability distribution has been also generalized into the resultant scalar measure of the uncertainty content in ψ:

$$S[\psi] = \int \psi^*(r) [-\ln p(r) - 2\phi(r)] \psi(r) \, dr \equiv \langle \psi | \hat{S} | \psi \rangle$$

$$= S[p] + S[\phi] \equiv \int p(\boldsymbol{r})\,[S_p(\boldsymbol{r}) + S_\phi(\boldsymbol{r})]\,d\boldsymbol{r} \equiv \int p(\boldsymbol{r})\,S(\boldsymbol{r})\,d\boldsymbol{r} \equiv \int \Sigma(\boldsymbol{r})\,d\boldsymbol{r}. \quad (2.31)$$

This expectation value of the (Hermitian) operator of the resultant global entropy,

$$\mathsf{S}\hat{}(\boldsymbol{r}) = -\,[\ln p(\boldsymbol{r}) + 2\phi(\boldsymbol{r})], \qquad (2.32)$$

combines the positive classical contribution $S[p]$ of Shannon and a negative nonclassical supplement $S[\phi] = -2\langle\phi\rangle$ reflecting the state average phase $\langle\phi\rangle = \int p(\boldsymbol{r})\,\phi(\boldsymbol{r})\,d\boldsymbol{r} \geq 0$.

To summarize, the Hermitian operator $\hat{I}(\boldsymbol{r})$ gives rise to real expectation value of the state resultant determinicity information content $I[\psi]$, related to the average kinetic energy $T[\psi]$, while the non-Hermitian entropy operator $\hat{S}(\boldsymbol{r})$ generates the complex average quantity $M[\psi]$. The classical and nonclassical densities-per-electron of the resultant gradient information and global entropy then separately obey the same, classical relations:

$$I_p(\boldsymbol{r}) = [\nabla M_p(\boldsymbol{r})]^2,\ I_\phi(\boldsymbol{r}) = [\nabla M_\phi(\boldsymbol{r})]^2 = [i\nabla S_\phi(\boldsymbol{r})]^2 = -\,[\nabla S_\phi(\boldsymbol{r})]^2, \quad (2.33)$$

and

$$\hat{I}(\boldsymbol{r}) = \nabla \hat{S}(\boldsymbol{r})^\dagger \cdot \nabla \hat{S}(\boldsymbol{r})^\dagger = |\nabla \hat{S}(\boldsymbol{r})|^2$$

$$= [\nabla \ln p(\boldsymbol{r})]^2 + [-2\nabla\phi(\boldsymbol{r})]^2 = [\nabla p(\boldsymbol{r})/p(\boldsymbol{r})]^2 + 4[\nabla\phi(\boldsymbol{r})]^2 \geq 0. \qquad (2.34)$$

Therefore, the gradient of complex entropy can be regarded as the quantum amplitude of the resultant information. In other words, $\nabla \hat{S}(\boldsymbol{r})$ appears as the "square root" of $\hat{I}(\boldsymbol{r})$. This development is thus in spirit of the quadratic approach of Prigogine.[17] It has been shown elsewhere[32] that the net productions of these resultant entropy/information quantities have exclusively nonclassical origins.

2.4   GENERAL ANALOGIES BETWEEN CLASSICAL AND QUANTUM DESCRIPTIONS

A comparison between the dynamical equations and their formal solutions indicates that a transition from the classical description to the quantum mechanical treatment of molecular states is effected by replacing the Poisson bracket $\{H, \rho\}$ with the commutator $[\hat{H}, \hat{\rho}]$, and the classical Liouville operator $\hat{\mathcal{L}}$ with its quantum analog $\hat{L}$.[17] In this mapping of the fundamental state variables, the Gibbs (real) probability density ρ of classical mechanics is associated with the quantum wavefunction (probability amplitude) ψ of the quantum description. One also observes that in formal solutions of dynamical equations describing the time–evolution of these state attribubes the Hamiltonian $\hat{H}$ replaces in QM the Liouville operator $\hat{\mathcal{L}}$ of the classical treatment. The non-Hermitian complex entropy operator of eq 2.25 formulated in terms of the electronic wavefunction, the fundamental quantum state "variable", thus fully conforms to these classical-quantum analogies.

The von Neuman's[44] entropy of the quantum ensemble, the mixed-state defined by the density operator $\hat{\rho}$, is determined by the mathematical trace

$$S_{vN}[\hat{\rho}] = \mathrm{tr}(\hat{\rho}\ln\hat{\rho}). \qquad (2.35)$$

Expressing $\hat{\rho}$ in terms of its eigenvectors $\{\psi_j\}$ and eigenvalues (probabilities) $\{|\eta_j\rangle\}$,

$$\hat{\rho}|\psi_i\rangle = \eta_i|\psi_i\rangle, \qquad (2.36)$$

$$\hat{\rho} = \sum_j |\psi_j\rangle\eta_j\langle\psi_j|, \qquad (2.37)$$

generates the information entropy contained in the ensemble probabilities:

$$S_{vN}[\hat{\rho}] = \sum_j \eta_j \ln\eta_j. \qquad (2.38)$$

This ensemble-average measure identically vanishes in the pure state $|\psi\rangle$, when $\hat{\rho}_\psi = |\psi\rangle\langle\psi|$ and $\eta_\psi = 1$: $S_{vN}[\hat{\rho}_\psi] = 0$.

The (idempotent) density operator $\hat{\rho}_\psi$ also determines the Hermitian density matrix in the position representation $\{|r\rangle\}$, $\langle r|r'\rangle = \delta(r'-r)$, represented by the kernel

$$\gamma_\psi(r,r') = \langle r|\hat{\rho}_\psi|r'\rangle = \psi(r)\,\psi^*(r'), \qquad \gamma_\psi(r,r) = p(r). \qquad (2.39)$$

In this *pure*-state case the "ensemble" average entropy,

$$S[\hat{\rho}_\psi] \equiv -\operatorname{tr}(\hat{\rho}_\psi \ln \hat{\rho}_\psi) = -\int dr \int dr'\, \gamma_\psi(r,r')\ln\gamma_\psi(r',r) \equiv S[\gamma_\psi], \qquad (2.40)$$

amounts to the quantum expectation value in $|\psi\rangle$ of the classical entropy operator $\hat{S}_\psi^{class.}$ defined by the diagonal kernel

$$\hat{S}_\psi^{class.}(r,r') = \langle r|\hat{S}_\psi^{class.}|r'\rangle = -\delta(r'-r)\ln p(r). \qquad (2.41)$$

It reduces to the classical Shannon entropy of the state probability density $p(r) = |\psi|^2$:

$$S[\hat{S}_\psi^{class.}] = \langle \psi|\hat{S}_\psi^{class.}|\psi\rangle = \int dr\, p(r)\ln p(r) = S[p]. \qquad (2.42)$$

The classical (Hermitian) entropy operator of eq 2.41 can be expressed in terms of the (non-Hermitian) resultant-entropy operators of eq 2.25:

$$\hat{S}_\psi^{class.} = \frac{1}{2}(\hat{S}_\psi + \hat{S}_\psi^*) = \operatorname{Re}(\hat{S}_\psi) = -\ln p = -[\ln\psi + \ln\psi^*] = 2\ln|\psi|. \quad (2.43)$$

Therefore, in the familiar Shannon entropy of the classical IT, which reconstructs the ensemble-average measure of von Neuman's quantum entropy in the *pure*-state density matrix, the phase/current information terms of complex entropies $S[\psi]$ and $S[\psi^*] = S[\psi]^*$ exactly cancel out, as indeed expected of the expectation value of the Hermitian operator $\hat{S}_\psi^{class.}$, which is devoid of any phase/current content.

One recalls that in QM one normally requires the observed physical properties to be represented by the associated (linear) Hermitian operators. However, the information entropy is neither observable, which can be directly determined in an experiment, nor is it linear in the underlying probability argument. Therefore, attributing to the overall quantum entropy content in the specified quantum state a non-Hermitian operator seems to be an admissible conceptual proposition, which is additionally capable of a unique phase characterization of the entangled molecular subsystems.[22,26] This attractive feature of the resultant IT concepts is particularly useful in their applications to reactive systems.[13,22]

Both the "real" approach of von Neumann and "complex" resultant approach are fully consistent with the classical-quantum analogies summarized at the beginning of this section. In this Hermitian development, exploring just the modulus part of the wavefunction, the classical Shannon entropy $S[p]$ is seen to be replaced by its quantum analog:

Classical: Quantum (Hermitian):

$$- \int d\mathbf{r}\, p(\mathbf{r}) \ln p(\mathbf{r}) \longleftrightarrow - \langle \psi | 2\ln|\psi| \, | \psi \rangle = - \langle \psi | \ln \psi + \ln \psi^* \, | \psi \rangle = - \langle \psi | \ln p | \psi \rangle = S[p].$$

In accordance with the same analogy, the full entropy content, probing both the *modulus* and *phase* components of the molecular wavefunction, calls for the non-Hermitian generalization of von Neumann's entropy:

Classical: Quantum (non-Hermitian):

$$- \int d\mathbf{r}\, p(\mathbf{r}) \ln p(\mathbf{r}) \longrightarrow - \langle \psi | 2\ln \psi | \psi \rangle = - \langle \psi | \hat{S}_\psi | \psi \rangle = S[p] + iS[\phi] = M[\psi].$$

2.5 CONCLUSION

In this short commentary, we have reexamined the resultant information/ entropy measures that combine the classical (probability) contributions of Fisher or Shannon and their associated nonclassical (phase/current) supplements. We have also explored general similarities between the classical and quantum mechanical descriptions. Both von Neumann's entropy in density matrix and the novel complex entropy conform to these analogies. The relation between the (complex) entropy of the resultant (non-Hermitian) approach and the (real) entropy of von Neumann's in the classical (Hermitian) IT treatment shows that the (imaginary) phase contribution of the former exactly cancels out in the latter. The classical description thus misses the crucial phase contribution, which is required for a full quantum IT characterization of the entangled molecular subsystems. The mutual relations between densities of such generalized Fisher (information) and Shannon (entropy) concepts have been summarized.

It has been stressed that their nonclassical (phase/current) components satisfy the same relations as do their classical (probability) contributions.

The overall IT descriptors allow one to distinguish the information content of states generating the same electron density but differing in their current composition. The electron density determines the "static" facet of molecular structure, the structure "of being," while the current distribution describes its "dynamic" aspect, the structure of "becoming." Both these aspects contribute to the overall information content of generally complex electronic states of molecular systems, reflected by the resultant IT quantities.

KEYWORDS

- **classical/quantum mechanics**
- **entropic descriptors**
- **information theory**
- **quantum electronic states**
- **resultant entropy/information**

REFERENCES

1. Fisher, R. A. Theory of Statistical Estimation. *Proc. Cambridge Phil. Soc.* **1925,** *22*, 700–725; See also: Frieden, B. R. Physics from the Fisher Information: A Unification; Cambridge University Press: Cambridge, 2004.

2. Shannon, C. E. The Mathematical Theory of Communication. *Bell System Tech. J.* **1948,** *27*, 379–493, 623–656; See also: Shannon, C. E.; Weaver, W. The Mathematical Theory of Communication. University of Illinois, Urbana, 1949.

3. Kullback, S.; Leibler, R. A. On Information and Sufficiency. *Ann. Math. Stat.* **1951,** *22*, 79–86; See also: Kullback, S. *Information Theory and Statistics*; Wiley: New York, 1959.

4. Abramson, N. *Information Theory and Coding.* McGraw-Hill: New York, 1963.

5. Pfeifer, P. E. *Concepts of Probability Theory*; Dover: New York, 1978.

6. Nalewajski, R. F.; Parr, R. G. Information Theory, Atoms-in-molecules and Molecular Similarity. *Proc. Natl. Acad. Sci. USA* **2000,** *97*, 8879–8882.

7. Parr, R. G.; Ayers, P. W.; Nalewajski, R. F. What Is an Atom in a Molecule? *J. Phys. Chem. A* **2005,** *109*, 3957–3959.

8. Nalewajski, R. F. Use of Fisher Information in Quantum Chemistry. *Int. J. Quantum Chem.* **2008,** *108*, 2230–2252.

9. Nalewajski, R. F. *Information Theory of Molecular Systems*; Elsevier: Amsterdam, 2006.

10. Nalewajski, R. F. *Information Origins of the Chemical Bond;* Nova Science Publishers: New York, 2010.

11. López-Rosa, S.; Esquivel, R. O.; Angulo, J. C.; Antolin, J.; Dehesa, J. S.; Flores-Gallegos, N. *J. Chem. Theory Comput.* **2010,** *6*, 145–154.

12. Nalewajski, R. F. *Perspectives in Electronic Structure Theory*; Springer: Heidelberg, 2012.

13. Nalewajski, R. F. *Quantum Information Theory of Molecular States;* Nova Science: Publishers, New York, 2016.

14. Zhou, X. Y.; Rong, C. Y.; Lu, T.; Zhou, P. P.; Liu, S. B. Information Functional Theory: Electronic Properties as Functionals of Information for Atoms and Molecules. *J. Phys. Chem. A* **2016,** *120*, 3634–3642.

15. Nalewajski, R. F. Electron Communications and Chemical Bonds. In *Frontiers of Quantum Chemistry*; Wójcik, M., Nakatsuji, H., Kirtman, B., Ozaki, Y., Eds.; Springer: Singapore, 2017; pp 315–351.

16. Heidar-Zadeh, F.; Ayers, P. W.; Verstraelen, T.; Vinogradov, I.; Vöhringer-Martinez, E.; Bultinck, P. Information-theoretic Approaches to Atoms-in-molecules: Hirshfeld Familty of Partitioning Schemes. *J. Phys. Chem. A* **2018,** *122*, 4219–4245.

17. Prigogine, I. *From Being to Becoming: Time and Complexity in the Physical Sciences*; Freeman WH & Co: San Francisco, 1980.

18. Nalewajski, R. F. On Phase/Current Components of Entropy/Information Descriptors of Molecular States; *Mol. Phys.* **2014,** *112*, 2587–2601.

19. Nalewajski, R. F. Quantum Information Descriptors in Position and Momentum Spaces. *J. Math. Chem.* **2015,** *53*, 1549–1575.

20. Nalewajski, R. F. Quantum Information Measures and Molecular Phase Equilibria. In *Advances in Mathematics Research*; Baswell, A. R., Ed.; Nova Science Publishers: New York, 2015; pp 53–86.

21. Nalewajski, R. F. Complex Entropy and Resultant Information Measures. *J. Math. Chem.* **2016,** *54*, 1777–1782.

22. Nalewajski, R. F. Phase Description of Reactive Systems. In *Conceptual Density Functional Theory and its Application in the Chemical Domain*; Islam, N., Kaya, S., Eds Apple Academic Press: Waretown, 2018; pp 217–249.

23. Nalewajski, R. F. Entropy Continuity, Electron Diffusion and Fragment Entanglement. In: *Equilibrium States in Advances in Mathematics Research*; Baswell, A. R., Ed.; Nova Science Publishers: New York, 2017, pp 1–42.

24. Nalewajski, R. F. Chemical Reactivity Description in Density-Functional and Information Theories. In *Chemical Concepts from Density Functional Theory*; Liu, S., Ed.; Acta Physico-Chimica Sinica, pp 2491–2509.

25. Primas, H. *Chemistry, Quantum Mechanics and Reductionism*; Springer-Verlag: Berlin, 1981.

26. Nalewajski, R. F. On Entangled States of Molecular Fragments. *Trends Phys. Chem.* **2016,** *16*, 71–85.

27. Nalewajski, R. F. Exploring Molecular Equilibria Using Quantum Information Measures. *Ann. Phys. (Leipzig)* **2013,** *525*, 256–268.

28. Nalewajski, R. F. On Phase Equilibria in Molecules. *J. Math. Chem.* **2014,** *52,* 588–612.
29. Nalewajski, R. F. Quantum Information Approach to Electronic Equilibria: Molecular Fragments and Elements of Non-equilibrium Thermodynamic Description. *J. Math. Chem.* **2014,** *52,* 1921–1948.
30. Nalewajski, R. F. Phase/Current Information Descriptors and Equilibrium States in Molecules. *Int. J. Quantum Chem.* **2015,** *115,* 1274–1288.
31. Nalewajski, R. F. On Entropy-continuity Descriptors in Molecular Equilibrium States. *J. Math. Chem.* **2016,** *54,* 932–954.
32. Nalewajski, R. F. Information Equilibria, Subsystem Entanglement and Dynamics of Overall Entropic Descriptors of Molecular Electronic Structure. *J. Mol. Model* (Chattaraj PK Issue), In Press, 2018.
33. Nalewajski, R. F. Entropic Measures of Bond Multiplicity from the Information Theory. *J. Phys. Chem. A* **2000,** *104,* 11940–11951.
34. Nalewajski, R. F. Multiple, Localized and Delocalized/Conjugated Bonds in the Orbital-communication Theory of Molecular Systems. *Adv. Quant. Chem.* **2009,** *56,* 217–250.
35. Nalewajski, R. F.; Szczepanik, D.; Mrozek, J. Bond Differentiation and Orbital Decoupling in the Orbital Communication Theory of the Chemical Bond. *Adv. Quant. Chem.* **2011,** *61,* 1–48.
36. Nalewajski, R. F. Entropy/Information Descriptors of the Chemical Bond Revisited. *J. Math. Chem.* **2011,** *49,* 2308–2329.
37. Nalewajski, R. F.; Köster, A. M.; Escalante, S. Electron Localization Function as Information Measure. *J. Phys. Chem. A* **2005,** *109,* 10038–10043.
38. Nalewajski, R. F.; de Silva, P.; Mrozek, J. Use of Non-additive Fisher Information in Probing the Chemical Bonds. *J. Mol. Struct. Theochem.* **2010,** *954,* 57–74.
39. Von Neumann, J. *Mathematical Foundations of Quantum Mechanics*; Prineton University Press: Princeton, 1955.
40. Harriman, J. E. Orthonormal Orbitals for the Representation of an Arbitrary Density. *Phys. Rev. A* **1981,** *24,* 680–682.
41. Zumbach, G.; Maschke, K. New Approach to the Calculation of Density Functionals. *Phys. Rev. A* **1983,** *28,* 544–554; Erratum *Phys. Rev. A* **1984,** *29,* 1585–1587.
42. Kohn, W.; Sham, L. J. Self-consistent Equations Including Exchange and Correlation Effects. *Phys. Rev.* **1965,** *140A,* 1133–1138.
43. Hohenberg, P.; Kohn, W. Inhomogeneous Electron Gas. *Phys. Rev.* **1964,** *136B,* 864–971.
44. von Weizsäcker, C. F. Zur theorie der kernmassen. *Z. Phys.* **1935,** *96,* 431–458.

CHAPTER 3

NONADDITIVE ENTROPIC CRITERIA FOR DENSITY PARTITIONING AND PHASE EQUILIBRIA

ROMAN F. NALEWAJSKI*

Department of Theoretical Chemistry, Jagiellonian University, Gronostajowa 2, 30-387 Cracow, Poland

E-mail: nalewajs@chemia.uj.edu.pl

ABSTRACT

The complex measure of the entropy (uncertainty) content in the specified molecular electronic state combines the classical (probability) contribution of Shannon and nonclassical supplement due to the state phase/current, as its real and imaginary parts, respectively. The associated resultant concept of the entropy deficiency in the given molecular state relative to the specified reference similarly reflects the overall information–resemblance between the two states compared. The condition of the vanishing nonadditive parts of these information–theoretic (IT) descriptors is shown to provide useful criteria for determining the mutual IT equilibria between subsystems: the real (classical) part determines the optimum density partitioning, while the imaginary (noclassical) contribution establishes the associated equilibrium phases. It is shown that the absolute entropy criterion of the mutual equilibrium between subsystems is satisfied by the Bader-type (physical space) partitioning of molecular electron density, while the vanishing relative-entropy requirement predicts the stockholder (functional-space) division of Hirshfeld. The latter criterion also determines the phase relations between subsystems, which are shown to be satisfied by the equilibrium ("thermodynamic") phases of molecular fragments.

3.1 INTRODUCTION

The classical information theory (IT)[1-8] has been successfully used to describe and interpret in chemical terms the electronic structure of molecules, for example, Refs. [9–12]. When applied to determine the optimum atomic fragments of molecular systems,[9–11,13–17] it rationalizes the Hirshfeld[18] "stockholder" partitioning of the molecular electron density. The information concepts also give insights[19] into the electron localization function[20–22] and generate the contragradience criterion[9–11,23] for localizing chemical interactions. The communication theory of the chemical bond[9–12,24–29] has provided a novel IT perspective on bond multiplicities, their covalent and ionic composition, and identified the intermediate (bridge) interactions in molecules.[30–35]

The electronic structure of molecules is embodied in their quantum states, which generate the system electron density and current distributions. A general electronic wavefunction is a complex entity characterized by its modulus and phase components. The square of the former determines the particle probability distribution, while the gradient of the latter generates the state current density. These complementary structural manifestations give rise to the associated classical and nonclassical contributions in the resultant measures[12,36–38] of the information/entropy content in complex electronic states. The overall entropy concept has been successfully applied to determine the molecular phase equilibria.[12,39–43] The (complex) resultant entropy,[12,37] the quantum expectation value of a non-Hermitian (state-dependent) entropy operator, generates in the resultant measure the probability and phase contributions as its real and imaginary parts. These generalized IT measures are also required to distinguish the mutually bonded (phase-related, "entangled") status of molecular fragments and reactants, from its nonbonded (phase-unrelated, "disentangled") counterparts.[43–46]

In this work, the (absolute) resultant entropy will be supplemented by the corresponding (relative) entropy deficiency concept, in spirit of the Kullback and Leibler development,[5,6] and the additive and nonadditive contributions to these generalized IT descriptors will be identified. The equilibrium criteria of the vanishing nonadditivities in density functionals for the overall resultant entropy and entropy deficiency will be shown to respectively generate the Bader-type[47] (physical-space) and Hirshfeld's[18] (functional-space) partitions of the molecular electron density into the corresponding atomic pieces.

3.2 ENTROPY AND ENTROPY DEFICIENCY DESCRIPTORS OF MOLECULAR STATES

We begin with some elementary reminders from molecular quantum mechanics. For simplicity, let us first consider a single-electron ($N = 1$) in state $|\psi\rangle$ at the specified time $t = t_0$, described by the associated (complex) wavefunction $\psi(r) = \langle r|\psi\rangle = R(r) \exp[i\phi(r)]$, where $R(r)$ and $\phi(r)$ stand for its modulus and phase parts. It generates the (normalized) particle probability density, $p(r) = \psi(r)^* \psi(r)$, $\int p(r)\,dr = 1$, and the current distribution

$$j(r) = [\hbar/(2mi)][\psi(r)^* \nabla \psi(r) - \psi(r) \nabla \psi(r)^*] = (\hbar/m)\, p(r) \nabla \phi(r) \equiv p(r)\, V(r), \tag{3.1}$$

where the velocity $V(r)$ of the probability fluid measures the current-per-particle:

$$V(r) = j(r)/p(r) = (\hbar/m) \nabla \phi(r). \tag{3.2}$$

The wavefunction modulus, a classical amplitude of the particle probability density, $p(r) = R(r)^2$, and the state phase, or its gradient determining the flux velocity, thus constitute the system fundamental degrees-of-freedom:

$$\psi \Leftrightarrow (R, \phi) \Leftrightarrow (p, j).$$

In molecules, the electron is moving in the external potential $v(r)$ due to the "frozen" nuclei. The electronic Hamiltonian,

$$\hat{H}(r) = - (\hbar^2/2m)\nabla^2 + v(r) \equiv \hat{T}(r) + v(r), \tag{3.3}$$

then determines the state dynamics, via the Schrödinger equation

$$i\hbar\, \partial \psi(\mathbf{r}, t)/\partial t = \hat{H}(\mathbf{r})\, \psi(\mathbf{r}, t), \tag{3.4}$$

which implies the sourceless continuity relation for the particle probability distribution $p(\mathbf{r}, t) = |\psi(\mathbf{r}, t)|^2$:

$$dp(\mathbf{r}, t)/dt = \partial p(\mathbf{r}, t)/\partial t + \nabla \cdot \mathbf{j}(\mathbf{r}, t) = 0. \tag{3.5}$$

The Shannon[3] descriptor of the classical global entropy in $p(\mathbf{r})$,

$$S[p] = -\int p(\mathbf{r})\ln p(\mathbf{r})\, d\mathbf{r} = -2\int R^2(\mathbf{r})\ln R(\mathbf{r})\, d\mathbf{r} \equiv S[R], \tag{3.6}$$

reflects the distribution "spread" (uncertainty), that is, a degree of the electron position "indeterminacy." It also provides the amount of information received, when this uncertainty is removed by the particle localization experiment. This classical measure generalizes into the corresponding resultant descriptor $S[\psi]$ of the quantum electronic state,[12,37] combining the probability ($S[p]$, real) and phase/current ($S[\phi]$, imaginary) components:

$$S[\psi] = \langle\psi|\hat{S}|\psi\rangle = S[p] - 2i\int p(\mathbf{r})\,\phi(\mathbf{r})\, d\mathbf{r} \equiv S[p] + iS[\phi]. \tag{3.7}$$

The complex average quantity represents the quantum expectation value of the state-dependent (non-Hermitian) entropy operator

$$\hat{S} = -2\ln\psi = -(\ln p + 2i\phi). \tag{3.8}$$

One similarly introduces the corresponding relative entropy (entropy deficiency) concepts describing the information distance between the two (isoelectronic) states:

$$\psi \Leftrightarrow (R, \phi) \Leftrightarrow (p, \mathbf{j}) \text{ and } \psi_0 \Leftrightarrow (R_0, \phi_0) \Leftrightarrow (p_0, \mathbf{j}_0). \tag{3.9}$$

The entropy deficiency (directed divergence) concept of Kullback and Leibler[5,6] measures the classical missing information in p relative to p_0:

$$\Delta S[p p_0] = \int p(\mathbf{r})\ln[p(\mathbf{r})/p_0(\mathbf{r})]\, d\mathbf{r}. \tag{3.10}$$

It measures the molecular average of the probability surprisal $S_p(r) = \ln[p(r)/p_0(r)]$. To avoid negative contributions, one also applies the unbiased Kullback's[6] measure of the probability divergence:

$$\Delta S[p, p_0] = \Delta S[p|p_0] + \Delta S[p_0|p] = \int [p(r) - p_0(r)]\ln[p(r)/p_0(r)]\, dr \geq 0. \quad (3.11)$$

These missing information descriptors provide measures of an information resemblance between the two compared probability distributions.[9,13]

The classical relative-entropy descriptor of eq 3.10 generalizes into the associated resultant concept suggested by eq 3.7,

$$\Delta S[\psi|\psi_0] = \Delta S[p|p_0] - 2i\int p(r)[\phi(r) - \phi_0(r)]\, dr \equiv \Delta S[p|p_0] + i\Delta S[\phi|\phi_0]. \quad (3.12)$$

It combines the classical (probability) functional of eq 3.10 and the phase supplement

$$\Delta S[\phi|\phi_0] = -2\int p(r)[\phi(r) - \phi_0(r)]\, dr \quad (3.13)$$

as the real and imaginary components, respectively. It should be observed that this generalized information distance identically vanishes when the two quantum states of eq 3.9 are identical. To avoid negative contributions to the average missing information measure for the equilibrium phase of the probability distribution,[12,39–43]

$$\phi_{eq.}[p] = -\tfrac{1}{2}\ln p, \quad (3.14)$$

one also introduces the unbiased nonclassical functional, the phase divergence defined by the arithmetic average

$$\Delta S[\phi, \phi_0] = \tfrac{1}{2}\{\Delta S[\phi|\phi_0] + \Delta S[\phi_0|\phi]\} = \int [p_0(r) - p(r)][\phi(r) - \phi_0(r)]\, dr \geq 0. \quad (3.15)$$

Alternative, logarithmic concepts of the relative-entropy contribution associated with the wavefunction phase, involving the phase-surprisal $S_\phi(r) = \ln[\phi(r)/\phi_0(r)]$, have been proposed elsewhere.[12]

Finally, by expressing the electron density $\rho(r) = Np(r)$ in terms of its shape function, the probability distribution $p(r)$, allows one to express the entropic density functionals in terms of the above *probability* concepts, for example,

$$S[\rho] = -\int\rho(r)\ln\rho(r)\,dr = NS[p] - N\ln N = NS[p] + const.,$$

$$\Delta S[\rho|\rho_0] = \int\rho(r)\ln[\rho(r)/\rho_0(r)]\,dr = N\Delta S[p|p_0],$$

$$\Delta M[\psi|\psi_0] = \Delta S[\rho|\rho_0] - 2i\int\rho(r)[\phi(r) - \phi_0(r)]\,dr \equiv N\{\Delta S[p|p_0] + i\Delta S[\phi|\phi_0]\}.$$
(3.16)

3.3   DENSITY PARTITION AND ADDITIVE/NONADDITIVE COMPONENTS OF ENTROPY FUNCTIONALS

Consider a division of the electron density $\rho(r)$ of a molecular system

$$R = [A|B] = A—B,$$
(3.17)

containing $N = \int\rho(r)dr$ electrons in the mutually-open (bonded, entangled) fragments $X = (A, B)$, into fragment distributions $\rho(r) = \{\rho_X(r)\}$:

$$\rho = \rho_A + \rho_B, \quad \int\rho_X\,dr = N_X, \quad N_A + N_B = N.$$
(3.18)

For example, these fragments can represent atoms-in-molecules (AIM) or their collections: functional groups, reactants, etc. This partition also implies the associated division of the probability (shape-factor) distribution $p(r) = \rho(r)/N$, $\int p(r)dr = 1$,

$$p = \rho_A/N + \rho_B/N = (N_A/N)(\rho_A/N_A) + (N_B/N)(\rho_B/N_B) \equiv P_A p_A + P_B p_B,$$
(3.19)

Here, the $p(r) = \{p_X(r)\}$ groups the fragment probability densities, unity normalized within each subsystem, and $P = \{P_X = N_X/N\}$ contains the corresponding condensed probabilities of molecular fragments in R:

$$\sum_X P_X = \int p_X(r)\,dr = 1.$$

It should be observed that the density pieces $\rho(r) = \{\rho_X(r) = N_X p_X(r)\}$ constitute true independent variables in the IT variational problems of density partitioning, since the optimized probability densities must be accompanied by the subsystem electron occupations $N = \{N_X\}$ to fully characterize their charge distributions. Therefore, the information principles for determining the optimum parts of ρ [eq 3.18] should be phrased in terms of density functionals, instead of probability functionals [see eq 3.16].

Consider first the resultant entropy contained in the molecular state ψ generating density ρ, $\psi \rightarrow \rho$,

$$M[\psi] \equiv -\int \rho(r)\ln\rho(r)\,dr - 2i\int \rho(r)\,\phi(r)\,dr$$

$$\equiv M[\rho] + i\,M[\phi] = N\{S[\rho] + iS[\phi]\} + const. \qquad (3.20)$$

It also represents the total resultant-entropy functional for the specified partitioning $\rho(r)$:

$$M^{\text{total}}[R] = -\sum_X \int \rho_X(r)[\ln\rho(r) + 2i\,\phi(r)]\,dr. \qquad (3.21)$$

Together with its additive part,

$$M^{\text{add.}}[R] = -\sum_X [\int \rho_X(r)\ln\rho_X(r)\,dr + 2i\int \rho_X(r)\,\phi_X(r)\,dr] \equiv \sum_X M_X[\psi], \qquad (3.22)$$

this overall functional thus determines the associated nonadditive contribution for the division in question:

$$M^{\text{nadd.}}[R] = M^{\text{total}}[R] - M^{\text{add.}}[R]$$

$$= -\sum_X \int \rho_X(r)\,\{\ln[\rho(r)/\rho_X(r)]\,dr + 2i\,[\phi(r) - \phi_X(r)]\}\,dr. \qquad (3.23)$$

This quantity measures a degree of a mutual entanglement ("overlap") of $\{\rho_X\}$ in ρ, thus reflecting the information independence of such density pieces.

In theories of chemical reactivity, one often invokes the promolecular reference:[12,18]

$$R^0 = [A^0|B^0] = A^0\text{-}B^0. \qquad (24)$$

It involves the "frozen," mutually-closed (nonbonded, disentangled) fragments $X^0 = (A^0, B^0)$, with the constituent atoms in their molecular positions. The promolecular state ψ_0, of the separated densities $\{\rho_X{}^0\}$, for example, the electron densities of the isolated constituent atoms, provides the chemically correct reference for interpreting the bond-formation processes. It generates the (isoelectronic) promolecular density ρ_0, $\psi_0 \rightarrow \rho_0$,

$$\rho_0 = \rho_A{}^0 + \rho_B{}^0, \qquad \rho_X{}^0 d\mathbf{r} = N_X{}^0, \qquad N_A{}^0 + N_B{}^0 = N^0 = N. \qquad (3.25)$$

Consider now a similar division of the relative resultant entropy descriptors. In order to connect to earlier (classical) derivations,[9,46] we use the directed divergence concepts of the preceding section. The determination of the nonadditive part of the resultant-entropy deficiency must involve the molecular entropy deficiency as the total descriptor of the density partition problem,

$$\Delta M[\psi|\psi_0] = \Delta S[\rho|\rho_0] + i\,\Delta S[\phi|\phi_0] =$$

$$= \sum_X \int \rho_X(\mathbf{r})\{\ln[\rho(\mathbf{r})/\rho_0(\mathbf{r})] - 2i\,[\phi(\mathbf{r}) - \phi_0(\mathbf{r})]\}\,d\mathbf{r} \equiv \Delta M^{\text{total}}[R|R^0], \quad (3.26)$$

and the associated additive contribution:

$$\Delta M^{\text{add.}}[R|R^0] = \sum_X \int \rho_X(\mathbf{r})\{\ln[\rho_X(\mathbf{r})/\rho_X{}^0(\mathbf{r})] - 2i\,[\phi_X(\mathbf{r}) - \phi_X{}^0(\mathbf{r})]\}\,d\mathbf{r}. \quad (3.27)$$

Together they generate the corresponding resultant measure of the nonadditivity in the entropy deficiency relative to promolecular reference:

$$\Delta M^{\text{nadd.}}[R|R^0] = \Delta M^{\text{total}}[R|R^0] - \Delta M^{\text{add.}}[R|R^0]$$

$$= \sum_X \int \rho_X \{\ln[(\rho/\rho_0)/(\rho_X/\rho_X{}^0)] - 2i\,[(\phi - \phi_0) - (\phi_X - \phi_X{}^0)]\}\,d\mathbf{r}. \quad (3.28)$$

3.4 NONADDITIVE ENTROPIES AS EQUILIBRIUM CRITERIA

It has been argued elsewhere that the classical nonadditive entropy deficiency can be used to rationalize the Hirshfeld[18] stockholder division of the molecular electron density.[9,46] In this section, we examine the associated equilibrium criteria implied by the nonadditive contributions to the resultant functionals discussed in the preceding section.

Let us first examine the equilibrium criterion of the vanishing nonadditive contribution to the absolute resultant entropy [see eq 3.23]:

$$M^{\text{nadd.}}[R] = 0. \qquad (3.29)$$

The classical (density) part of this functional vanishes, when $\rho_X{}^B(\mathbf{r})$ = $\rho(\mathbf{r})$, X = A, B, that is, for the physical-space division of the molecular density, with subsystems defined by the density pieces corresponding to the exclusive domains in the physical space, for example, those obtained in the topological partitioning of Bader (B).[47] Accordingly, the nonclassical contribution to the nonadditive resultant entropy vanishes when the phases of subsystems equalize with the molecular phase, $\phi_X(r) = \phi(r)$, as indeed expected of the mutually open (bonded, entangled) constituent fragments of R.[42]. The vanishing nonadditivity of the resultant entropy should indeed be expected in the non-overlapping equilibrium densities of such spatial molecular fragments, which result from the physical-space division.

Next, let us examine the equilibrium division resulting from the relative resultant entropy criterion of the vanishing nonadditive relative resultant entropy [see eq 3.28]:

$$\Delta M^{\text{nadd.}}[R|R^0] = 0. \tag{3.30}$$

Its real (classical) part is seen to allow for the optimum function-space partition of the molecular electron density, into the overlapping fragments. It vanishes when the local density enhancement for the system as a whole, $w^H(r) \equiv \rho(r)/\rho_0(r)$, also characterizes each molecular fragment:

$$w_X(r) \equiv \rho_X(r)/\rho_X^0(r) = \rho(r)/\rho_0(r) = w^H(r), \qquad X = A, B. \tag{3.31}$$

This is the characteristic feature of the stockholder division scheme of Hirshfeld (H),[18] $w_X^H(r) = w^H(r)$:

$$\rho_X^H = {}_X^0 w^H \equiv d_X^0 \text{ or } d_X^0 \equiv \rho_X^0/\rho^0 = \rho_X^H/\rho \equiv d_X^H. \tag{3.32}$$

These optimum atomic pieces of the molecular density can be thus regarded as either the molecularly enhanced densities $\{\rho_X^0\}$ of the isolated atoms, or as the promolecular shares $\{d_X^0(r) = \rho_X^0(r)/\rho^0(r)\}$ of the molecular density $\rho(r)$; here, the promolecular distribution $\rho^0(r) = \sum_X \rho_X^0(r)$ is determined by densities of free atoms shifted to their positions in the molecule.

The vanishing nonclassical part of the resultant entropy deficiency also predicts the phase relation

$$\phi_X(r) - \phi_X^0(r) = \phi(r) - \phi_0(r), \qquad X = A, B. \tag{3.33}$$

It can be directly verified that the equilibrium "thermodynamic" phase of eq 3.14 satisfies this coherence requirement. Indeed, for the stockholder densities of eq 3.30 one finds

$$\phi_X(r) - \phi_X^0(r) = \phi(r) - \phi_0(r) = -\frac{1}{2} \ln w^H(r). \tag{3.34}$$

Therefore, the criterion of the vanishing nonadditivity of the resultant entropy deficiency rationalizes the stockholder fragments, resulting as the optimum, equilibrium pieces of the molecular electron density. It also properly identifies the relation satisfied by the subsystem equilibrium phases.[12,39–43]

3.5   CONCLUSION

This analysis stresses further a need for the nonclassical (phase/current) supplements of the classical (probability) contributions in the resultant entropic measures of the information content of the molecular electronic states. Such generalized concepts allow one to distinguish the information content of states generating the same electron density but differing in their current composition. They also facilitate an IT description of the bonding (entanglement) status of molecular subsystems. The phase aspect of molecular states is also vital for the quantum (amplitude) communications between AIM, which determine entropic descriptors of the chemical bond multiplicities and their covalent/ionic components.

In the DFT-based theory of chemical reactivity, one distinguishes between several hypothetical stages involving either the mutually open (bonded, entangled) or the mutually closed (nonbonded, disentangled) states of reactants, for the same electron distribution in the constituent subsystems. These two categories are discerned by the phase aspect of the quantum entanglement between reactants. This generalized approach deepens our understanding of the molecular/promolecular promotions of the constituent molecular fragments and provides a more precise framework for describing the hypothetical stages invoked in the theory of the chemical bond and reactivity.

This analysis has examined nonadditivities of the quantum entropy and entropy deficiency concepts, which arise in the partition of the molecular electron density. Such functionals can be used to formulate the IT principles for determining the equilibria in molecular subsystems, the division of electronic density into the least-dependent fragments, etc. The densities of the optimum subsystems and the phase relations between them both result from the equilibrium criterion of the vanishing resultant entropic nonadditivity. The nonadditivity in the resultant entropy was shown to vanish for the physical-space division of Bader, into the nonoverlaping subsystems, with the fragment phases being determined by the molecule as a whole. The corresponding IT criterion of the vanishing resultant entropy deficiency predicts the function-space division of Hirshfeld, into the overlapping stockholder pieces of molecular density. These IT optimum densities have been shown to also obey the predicted equilibrium-phase relations.

KEYWORDS

- complex entropy
- density partitioning
- equilibrium principles
- information theory
- phases in subsystems
- resultant entropy deficiency

REFERENCES

1. Fisher, R. A. Theory of Statistical Estimation. *Proc. Cambridge Phil. Soc.* **1925,** *22,* 700–725.
2. Frieden, B. R. *Physics from the Fisher Information: A Unification.* Cambridge University Press: Cambridge, 2004.
3. Shannon, C. E. The Mathematical Theory of Communication. *Bell System Tech. J.* **1948,** *27,* 379–493, 623–656.
4. Shannon, C. E.; Weaver, W. *The Mathematical Theory of Communication*; University of Illinois: Urbana, 1949.
5. Kullback, S.; Leibler, R. A. On Information and Sufficiency. *Ann. Math. Stat.* **1951,** *22,* 79–86.
6. Kullback, S. *Information Theory and Statistics*; Wiley: New York, 1959.
7. Abramson, N. *Information Theory and Coding*; McGraw-Hill: New York, 1963.
8. Pfeifer, P. E. *Concepts of Probability Theory*; Dover: New York, 1978.
9. Nalewajski, R. F. *Information Theory of Molecular Systems*; Elsevier: Amsterdam, 2006.

10. Nalewajski, R. F. *Information Origins of the Chemical Bond*; Nova Science: Publishers, New York, 2010.
11. Nalewajski R. F. *Perspectives in Electronic Structure Theory*; Springer: Heidelberg, 2012.
12. Nalewajski, R. F. *Quantum Information Theory of Molecular States*; Nova Science Publishers: New York, 2016.
13. Nalewajski, R. F.; Parr, R. G. Information Theory, Atoms-in-molecules and Molecular Similarity. *Proc. Natl. Acad. Sci. USA* **2000**, *97*, 8879–8882.
14. Nalewajski, R. F.; Parr, R. G. Information-theoretic Thermodynamics of Molecules and their Hirshfeld Fragments. *J. Phys. Chem. A* **2001**, *105*, 7391–7400.
15. Nalewajski, R. F. Hirschfeld Analysis of Molecular Densities: Subsystem Probabilities and Charge Sensitivities. *Phys. Chem. Chem. Phys. 4* **2002**, 1710–1721.
16. Parr, R. G.; Ayers, P. W.; Nalewajski, R. F. What Is an Atom in a Molecule? *J. Phys. Chem. A* **2005**, *109*, 3957–3959.
17. Heidar-Zadeh, F.; Ayers, P. W.; Verstraelen, T.; Vinogradov, I.; Vöhringer-Martinez, E.; Bultinck, P. Information-theoretic Approaches to Atoms-in-molecules: Hirshfeld Familty of Partitioning Schemes. *J. Phys. Chem. A* **2018**, *122*, 4219–4245.
18. Hirshfeld, F. L. Bonded-atom Fragments for Describing Molecular Charge Densities. *Theor. Chim. Acta.* (Berl) **1997**, *44*, 129–138.
19. Nalewajski, R. F.; Köster, A. M.; Escalante, S. Electron Localization Function as Information Measure. *J. Phys. Chem. A* **2005**, *109*, 10038–10043.
20. Becke, A. D.; Edgecombe, K. E. A Simple Measure of Electron Localization in Atomic and Molecular Systems. *J. Chem. Phys.* **1990**, *92*, 5397–5403.
21. Silve, B.; Savin, A. Classification of Chemical Bonds Based on Topological Analysis of Electron Localization Functions. *Nature* **1994**, *371*, 683–686.
22. Savin, A.; Nesper, R.; Wengert, S.; Fässler, T. F. ELF: The Electron Localization Function. *Angew. Chem. Int. Ed. Engl.* **1997**, *36*, 1808–1832.
23. Nalewajski, R. F.; de Silva, P.; Mrozek, J. Use of Nonadditive Fisher Information in Probing the Chemical Bonds. *J. Mol. Struct. Theochem.* **2010**, *954*, 57–74.
24. Nalewajski, R. F. Entropic Measures of Bond Multiplicity from the Information Theory. *J. Phys. Chem. A* **2000**, *104*, 11940–11951.
25. Nalewajski, R. F. Multiple, Localized and Delocalized/Conjugated Bonds in the Orbital-communication Theory of Molecular Systems. *Adv. Quant. Chem.* **2009**, *56*, 217–250.
26. Nalewajski, R. F.; Szczepanik, D.; Mrozek, J. Bond Differentiation and Orbital Decoupling in the Orbital Communication Theory of the Chemical Bond. *Adv. Quant. Chem.* **2011**, *61*, 1–48; *Ibid.* Basis Set Dependence of Molecular Information Channels and Their Entropic Bond Descriptors. *J. Math. Chem.* **2012**, *50*, 1437–1457.
27. Nalewajski, R. F. Electron Communications and Chemical Bonds. In *Frontiers of Quantum Chemistry*; Wójcik, M., Nakatsuji, H., Kirtman, B., Ozaki, Y., Eds.; Springer: Singapore, 2017; pp 315–351.
28. Nalewajski, R. F. Entropy/Information Descriptors of the Chemical Bond Revisited. *J. Math. Chem.* **2011**, *49*, 2308–2329.
29. Nalewajski, R. F. Quantum Information Descriptors and Communications in Molecules. *J. Math. Chem.* **2014**, *52*, 1292–1323.
30. Nalewajski, R. F. Through-space and Through-bridge Components of Chemical Bonds. *J. Math. Chem.* **2011**, *49*, 371–392.

31. Nalewajski, R. F. Chemical Bonds from Through-bridge Orbital Communications in Prototype Molecular Systems. *J. Math. Chem.* **2011**, *49*, 546–561.

32. Nalewajski, R. F. On Interference of Orbital Communications in Molecular Systems. *J. Math. Chem.* **2011**, *49*, 806–815.

33. Nalewajski, R. F.; Gurdek, P. On the Implicit Bond-dependency Origins of Bridge Interactions. *J. Math Chem.* **2011**, *49*, 1226–1237.

34. Nalewajski, R. F. Direct (Through-space) and Indirect (Through-bridge) Components of Molecular Bond Multiplicities. *Int. J. Quantum Chem. 112*, 2355–2370.

35. Nalewajski, R. F.; Gurdek, P. Bond-order and Entropic Probes of the Chemical Bonds. *Struct. Chem.* **2012**, *23*, 1383–1398.

36. Nalewajski, R. F. On Phase/Current Components of Entropy/Information Descriptors of Molecular States. *Mol. Phys.* **2014**, *112*, 2587–2601.

37. Nalewajski, R. F. Complex Entropy and Resultant Information Measures. *J. Math. Chem.* **2016**, *54*, 1777–1782.

38. Nalewajski, R. F. Quantum Information Measures and Their Use in Chemistry. *Curr. Phys. Chem.* **2017**, *7*, 94–117.

39. Nalewajski, R. F. Exploring Molecular Equilibria Using Quantum Information Measures. *Ann. Phys.* (Leipzig) **2013**, *525*, 256–268.

40. Nalewajski R. F. On Phase Equilibria in Molecules. *J. Math. Chem.* **2014**, *52*, 588–612.

41. Nalewajski, R. F. Quantum Information Approach to Electronic Equilibria: Molecular Fragments and Elements of Non-equilibrium Thermodynamic Description. *J. Math Chem.* **2014**, *52*, 1921–1948.

42. Nalewajski, R. F. Phase/Current Information Descriptors and Equilibrium States in Molecules. *Int. J. Quantum Chem.* **2015**, *115*, 1274–1288.

43. Nalewajski, R. F. Quantum Information Measures And Molecular Phase Equilibria. In *Advances in Mathematics Research*, Baswell, A. R., Ed.; Nova Science Publishers: New York, 2015, pp 53–86.

44. Nalewajski, R. F. Phase Description of Reactive Systems. In *Conceptual Density Functional Theory*; Islam, N., Kaya. S., Eds.; Apple Academic Press: Waretown, pp 217–249.

45. Nalewajski, R. F. Entropy Continuity, Electron Diffusion and Fragment Entanglement. In: *Equilibrium States in Advances in Mathematics Research,* Baswell, A. R., Ed.; Nova Science Publishers: New York, 2017; pp 1–42.

46. Nalewajski, R. F. On Entangled States of Molecular Fragments. *Trends Phys. Chem.* **2016**, *16*, 71–85.

47. 45. Nalewajski, R. F. Chemical Reactivity Description in Density-functional and Information Theories. In *Chemical Concepts from Density Functional Theory*; Liu, S., Ed; Acta Physico-Chimica Sinica, 2017; pp 2491–2509.

48. Nalewajski, R. F. Information Equilibria, Subsystem Entanglement and Dynamics of Overall Entropic Descriptors of Molecular Electronic Structure. *J. Mol. Model* (Chattaraj PK Issue), In Press, 2018.

49. Bader, R. F. W. *Atoms in Molecules*. Oxford University Press: New York, 1990.

CHAPTER 4

EFFECT OF CHEMICAL REACTIONS ON PRESSURE FLUCTUATIONS IN THE COMBUSTION CHAMBER

A. V. ALIEV and O. A. VOEVODINA[*]

Federal State Budgetary Educational Institution of Higher Education, Kalashnikov Izhevsk State Technical University, Izhevsk Studencheskaya St. 7, Russia, 426069

[*]*Corresponding author. E-mail: voevodina.o.a@yandex.ru*

ABSTRACT

The reasons for oscillations of intraballistic parameters in the solid fuel engine chamber are considered. The influence of acoustic processes is noted and examples of their evaluation are given. Along with acoustic processes, it is proposed to evaluate the effect of chemical reactions on pressure fluctuations in the combustion chamber. To solve the problem, along with the mathematical model of thermodynamic processes for chemically reacting combustion products, its linearized version is considered. The eigenvalues of the matrix of the linearized model are investigated, which allows one to draw conclusions about the presence or absence of a frequency spectrum due to chemical reactions.

4.1 INTRODUCTION

The operation of the solid propellant rocket engine on nonstationary and quasi-stationary operation modes is accompanied by influence of perturbing factors on intraballistic parameters.[1-3] Perturbing factors can be periodic (fluctuating) or nonperiodic. The cause of perturbations can be the physical characteristics of solid fuel combustion (e.g., the discreteness of oxidant or fuel burnup in the blended fuel). A description of the main regularities of this type of disturbance in the engine chamber is contained, for example, in Refs. [4–6]. Acoustic phenomena due to multiple reflections of pressure waves from the walls of the chamber of the solid propellant rocket engine at all stages of its operation, the formation and movement of stable vortex structures along the volume of the chamber, is an object of study in the works of the last century (e.g., Refs. 7–9), years (e.g., Refs. 10–13). Perturbing factors affecting the value of the pressure in the chamber of solid propellant rocket engine can arise due to chemical reactions occurring between the individual components of the combustion products. An analysis of such phenomena was carried out, for example, in Refs. [14–16].

The solution of the problem of stability of intrachamber processes and the determination of vibration frequencies in the case of their occurrence in the majority of the listed works is based on methods for analyzing the stability of dynamical systems. In Ref. [17], when analyzing the perturbations that arise in the rocket engine of solid fuel of the channel type at the initial stage of its operation (Fig. 4.1a), the characteristic equations were used which allowed setting the own frequencies of the propulsion system at an arbitrary time. Characteristics equations are obtained by linearizing the nonstationary one-dimensional equations of internal ballistics of a solid propellant rocket. The coefficients of the matrix from which its eigenvalues are determined were determined from the results of the solution of the task of the flow of combustion products of solid propellant in the charge channel of the solid propellant. The approach used in Ref. [17] made it possible to establish for each section of the combustion chamber at any time the amplitude and frequency of natural oscillations (if they exist).[17]

It was shown in Ref. [17] that the oscillations arising in the combustion chamber of the rocket engine of solid fuel are a function of time and the spatial coordinate in question. Figure 4.2 shows the values of the oscillation frequencies in the central channel of the propellant rocket propellant (Fig. 4.1a) for various sections (coordinates 0.5, 2.4, and 5.8 m, respectively, measured from the front of the engine).

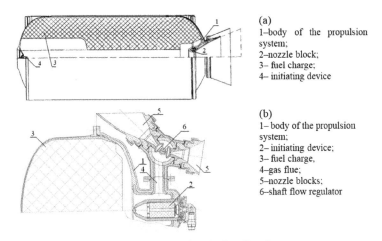

(a)
1–body of the propulsion system;
2–nozzle block;
3– fuel charge;
4– initiating device

(b)
1– body of the propulsion system;
2– initiating device;
3– fuel charge,
4–gas flue;
5–nozzle blocks;
6–shaft flow regulator

(a) sustainer engine; (b) adjustable engine

FIGURE 4.1 Constructive diagrams of the rocket engine of solid fuel.

As follows from Figure 4.2, one of the dangerous frequencies of oscillation is the frequency ~ 72 Hz, which is manifested in the vicinity of the ignition system at the time ~0.2–0.25 s. The conclusions[17] were subsequently confirmed experimentally in the field tests of a large rocket engine of solid fuel.

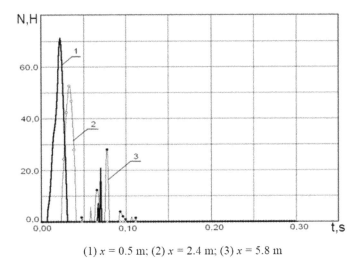

(1) $x = 0.5$ m; (2) $x = 2.4$ m; (3) $x = 5.8$ m

FIGURE 4.2 Spectrum of frequencies in the central channel of the fuel charge.

Below we will consider intrachamber processes in solid propellant controlled propulsion system, the constructive scheme of which is shown in Figure 4.1b. Calculation of intraballistic parameters in the solid propellant controlled propulsion system will be performed in the formulation applied in Ref. [18]:

- thermogasdynamic parameters are determined in the averaged over the volume of the combustion chamber;
- the composition of the combustion products at each step in time is established on the assumption of chemical equilibrium in the engine chamber, determined by the values of pressure and temperature;
- when solving the problem of the development of perturbations of thermogasdynamic parameters in the chamber of a solid propellant rocket engine, the approach used in Ref. [17] will be used.

When writing equations describing intracametric processes, the following basic notations are used:

- t—process time
- ρ_m, u_m, S_m—the fuel density, the burning rate, the combustion surface area of the fuel charge;
- ρ, p, T, E, H—density, pressure, temperature, internal energy and total heat content of products in the gas phase;
- α–the mass concentration of individual components in combustion products;
- c_p, c_v, R, k–the values of specific heats for combustion products, gas constant, and ratio of specific heats;
- 0, a ,к—indices denoting initial values of parameters, values in the environment, and parameters in the combustion chamber;
- A_i, $i = 1$, N–chemical compounds in the composition of combustion products;
- W, F_{min}, m—the volume of the combustion chamber, the area of the minimum section of the nozzle block, the flow coefficient;
- G_m, G_c—mass entry into the chamber from burning fuel and a second mass flow of combustion products from the chamber.

The system of equations providing determination of the intraballistic characteristics of a solid rocket engine of control after end of work of system of ignition is averaged over the volume of the combustion chamber statement which can be written in the form:

- the equation for changing the free volume of the combustion chamber of a solid propellant rocket engine of control and the thermodynamic equations of conservation of mass and energy in the volume (taking into account the loss of heat to the open surface of the body of the propulsion system S_1)

$$\frac{dW}{dt} = S_m u_m ,$$

(4.1)

$$\frac{d\rho W}{dt} = G_m - G_c ,$$

(4.2)

$$\frac{d\rho WE}{dt} = G_{6c} H_6 + G_m H_m - kG_c E - q_M S_M ;$$

(4.3)

equation of state and equations for thermophysical and thermodynamic parameters

$$p = \rho(k-1) E, T = \frac{E}{c_v} .$$

(4.4)

equations for the arrival G_m and consumption G_c of combustion products from the chamber

$$G_m = \rho_m u_m S_m ,$$

(4.5)

$$G_c = A_c p\mu F_{min} ,$$

(4.6)

$$A_c = \begin{cases} \sqrt{\dfrac{k}{RT}\left(\dfrac{2}{k+1}\right)^{\frac{k+1}{2(k-1)}}} & \text{при } \dfrac{p_a}{p_\kappa} \leq \left(\dfrac{2}{k+1}\right)^{\frac{k}{k-1}} , \\[3em] \sqrt{\dfrac{2k}{(k-1)RT}\left[\left(\dfrac{p_a}{p_\kappa}\right)^{\frac{2}{k}} - \left(\dfrac{p_a}{p_\kappa}\right)^{\frac{k+1}{k}}\right]} & \text{при } \left(\dfrac{2}{k+1}\right)^{\frac{k}{k+1}} < \dfrac{p_a}{p_\kappa} < 1, \\[3em] 0 & \text{при } \dfrac{p_a}{p_\kappa} > 1; \end{cases}$$

(4.7)

Equations of thermodynamic processes in the chamber of the propulsion system should be supplemented with equations for changing the mass concentrations α_i ($I = 1, N$), which is due to the chemical reactions occurring in the chamber of the propulsion system.

We will assume that the combustion products consist of N component A_i, hose mass concentration α_i as a function of the pressure and temperature of the combustion products in the chamber $\alpha_i(p,T)$ due to the chemical reactions of the type

$$qA_i + rA_j \Leftrightarrow sA_k + tA_l \tag{4.8}$$

Here i, j, k, l—the serial number of the component that takes a value in the range from 1 to N (it is assumed that the listed component numbers do not match); q, r, s, t—number of moles of reagents taking part in the reaction. The recorded reaction can proceed in both directions (direct and reverse reactions are realized).

The rates of direct K' and reverse K'' reactions are determined by the Arrhenius laws

$$K' = k' \cdot T^n \cdot \exp\left(-\frac{C}{T}\right),$$

$$K'' = k'' \cdot T^m \cdot \exp\left(-\frac{D}{T}\right).$$

Values $K', K'', k', k'', n, m, C, D$, included in the above equations are established experimentally or using semi-empirical approaches.

For reactions (4.8), the change in the mass concentrations of the components participating in the reaction can be established by equations

$$\frac{d\rho\alpha_i W}{dt} = G_m\alpha_{i0} - G_c\alpha_i + \rho W \cdot \left(K''\alpha_k^s\alpha_l^t - K'\alpha_i^q\alpha_j^r\right),$$

$$\frac{d\rho\alpha_j W}{dt} = G_m\alpha_{j0} - G_c\alpha_j + \rho W \cdot \left(K''\alpha_k^s\alpha_l^t - K'\alpha_i^q\alpha_j^r\right),$$

$$\frac{d\rho\alpha_k W}{dt} = G_m\alpha_{k0} - G_c\alpha_k + \rho W \cdot \left(K'\alpha_i^q\alpha_j^r - K''\alpha_k^s\alpha_l^t\right),$$

$$\frac{d\rho\alpha_i W}{dt} = G_m\alpha_{i0} - G_c\alpha_i + \rho W \cdot \left(K'\alpha_i^q\alpha_j^r - K''\alpha_k^s\alpha_l^t \right)$$

In reality, several chemical reactions are realized in combustion products, leading to a change in the concentration of the component ($i = 1, N$). For convenience, we rewrite eq 4.8 for the j-th reaction in the form

$$a_{j1}A_1 + a_{j2}A_2 + \ldots + a_{jN}A_N \Leftrightarrow b_{j1}A_1 + b_{j2}A_2 + \ldots + b_{jN}A_N$$

We note that among the coefficients a_{j*} and b_{j*} only two coefficients (respectively, in the left and right sides of the equation) are nonzero (bimolecular reactions are considered). The rate of direct and reverse reactions will be denoted, respectively K'_j, K''_j.

We denote by

$$\Psi'_j = K'_j \cdot \alpha_1^{a_{j1}}\alpha_2^{a_{j2}} \cdot \ldots \cdot \alpha_N^{a_{jN}},$$

$$\Psi''_j = K''_j \cdot \alpha_1^{b_{j1}}\alpha_2^{b_{j2}} \cdot \ldots \cdot \alpha_N^{b_{jN}}$$

$$\delta_{ij} = \begin{cases} -1, \text{ if in the } j-\text{th direct reaction the value } \alpha_i \text{ decreases,} \\ 1, \text{ if in the } j-\text{th direct reaction the value } \alpha_i \text{ increases,} \\ 0, \text{ if } j-\text{th reaction does not change the value of the } \alpha_i \end{cases}$$

With the notations written down, the equation for changing the concentration α_i can be written in the form

$$\frac{d\rho\alpha_i W}{dt} = G_m\alpha_{i0} - G_c\alpha_i + \rho W \sum_{j=1}^{M} \delta_{ij} \left(\Psi'_j - \Psi''_j \right)$$

or taking into account eq 4.2

$$\frac{d\alpha_i}{dt} = \frac{G_m(\alpha_{i0} - \alpha_i)}{\rho W} + \sum_{j=1}^{M} \delta_{ij} \left(\Psi'_j - \Psi''_j \right) \tag{4.9}$$

The thermophysical characteristics of the combustion products are functions of mass concentrations α_i ($i = 1, N$); therefore, the system of equations for determining the thermogasdynamic parameters in the combustion chamber should be supplemented by the relations

$$c_p = \sum_{i=1}^{N} c_{pi}\alpha_i, \quad c_v = \sum_{i=1}^{N} c_{vi}\alpha_i, \quad R = c_p - c_v, \quad k = {c_p}\big/{c_v}.$$

If the solution of the problem is considered in a neighborhood $\frac{d\alpha_i}{dt} \approx 0$, it is possible to apply the linearization procedure of eq 4.9. The matrix, which makes it possible to estimate the stability of the quasi-stationary operating mode of the propulsion system, due to chemical reactions, and which makes it possible to determine the frequencies of the natural vibrations (if any), according to Refs. [19, 20], can be written in the form

$$J = \begin{Bmatrix} a_{11} & a_{12} & \cdots & a_{1j} & \cdots & a_{1N} \\ a_{21} & a_{22} & \cdots & a_{2j} & \cdots & a_{2N} \\ \cdots & \cdots & \cdots & \cdots & \cdots & \cdots \\ a_{i1} & a_{i2} & \cdots & a_{ij} & \cdots & a_{iN} \\ \cdots & \cdots & \cdots & \cdots & \cdots & \cdots \\ a_{N1} & a_{N2} & \cdots & a_{Nj} & \cdots & a_{NN} \end{Bmatrix}$$

Here $a_{ij} = \begin{cases} -\dfrac{G_m}{\rho W} + \dfrac{1}{\alpha_j} \sum_{k=1}^{N} \delta_{ik}\left(a_{kj}\Psi'_k - b_{kj}\Psi''_k\right), & \text{for } i = j; \\ \dfrac{1}{\alpha_j} \sum_{k=1}^{N} \delta_{ik}\left(a_{kj}\Psi'_k - b_{kj}\Psi''_k\right), & \text{for } i \neq j. \end{cases}$

4.2 EXPERIMENTAL

As an example, we perform an analysis of the stability of the solid propellant rocket engine of control (Fig. 4.1b), caused by perturbations arising from the occurrence of chemical reactions between the components of chemical reactions. The analysis is feasible with the following initial data:

- conditional fuel formula—$C_{16.7}H_{40.77}O_{25.1}N_{17.8}Cl_{3.0}$;
- pressure of combustion chamber—10.0 MPa;
- temperature of combustion products—1520 K.

The most significant chemical compounds which are formed when burning fuel, their mass concentration in combustion products, and values derivative of mass concentration on pressure are given in Table 4.1. It is supposed that combustion chamber pressure decreases from 10.0 MPa to 9.0 MPa. The analysis of the results provided in Table 4.1 shows that the most significant responses are such as a result of which concentration of the chemical compounds placed in the first seven lines of Table 4.1 change (concentration of the components located below the seventh line practically do not change).

The list of the main chemical reactions proceeding when burning fuel and at which course concentration of the most significant chemical compounds change is provided in Table 4.2. The constants determining speeds of direct and return chemical reactions are given in the same table. The analysis of the reactions presented in Table 4.2 shows that change of concentration of the connection CO (α_1) results from reactions with numbers 9–23 and 25. Change of concentration of hydrogen H_2 (α_2) results from reactions with numbers 1, 15, 27, and 31. The 25th, 26th, and 28th reactions exert impact on change of concentration of nitrogen N_2 (α_3). Reactions 1–8, 16 change concentration of vapors of water H_2O (α_4) in combustion products. Reactions 10, 19–22, 25 change concentration of carbon dioxide CO_2(α_5). Concentration of hydrochloric acid HCl (α_6) changes at course of chemical reaction with number 30. Concentration of methane CH_4 (α_7) changes as a result of reactions with numbers 14, 24, and 29.

TABLE 4.1 Composition of Combustion Products and the Dependence of the Concentrations on the Pressure in the Combustion Chamber.

No. p/p	Chemical compound	α_i	$d\alpha_i \big/ dp$, $\dfrac{1}{MPa}$
1	H_2	$2.40842920 \cdot 10^{-1}$	$2.81039000 \cdot 10^{-3}$
2	CH_4	$2.38612330 \cdot 10^{-2}$	$1.18929500 \cdot 10^{-3}$
3	H_2O	$1.27401940 \cdot 10^{-1}$	$1.33240000 \cdot 10^{-3}$
4	CO	$2.62857950 \cdot 10^{-1}$	$6.29860000 \cdot 10^{-4}$
5	N_2	$1.96169480 \cdot 10^{-1}$	$4.33430000 \cdot 10^{-4}$
6	CO_2	$8.19557040 \cdot 10^{-2}$	$2.88988000 \cdot 10^{-4}$
7	HCl	$6.62349530 \cdot 10^{-2}$	$1.52963000 \cdot 10^{-4}$
8	NH_3	$6.39343820 \cdot 10^{-4}$	$3.98648400 \cdot 10^{-5}$

TABLE 4.1 *(Continued)*

No. p/p	Chemical compound	α_i	$d\alpha_i / dp$, $\frac{1}{MPa}$
9	HCN	$2.47315660 \cdot 10^{-5}$	$1.87291100 \cdot 10^{-6}$
10	C_2H_6	$2.88093620 \cdot 10^{-6}$	$4.53553500 \cdot 10^{-7}$
11	H_2CO	$2.80738350 \cdot 10^{-6}$	$2.59892000 \cdot 10^{-7}$
12	C_2H_4	$2.10661170 \cdot 10^{-6}$	$2.80260900 \cdot 10^{-7}$
13	H_2CO_2	$2.03314320 \cdot 10^{-6}$	$2.20732400 \cdot 10^{-7}$
14	CH_3Cl	$1.49405590 \cdot 10^{-6}$	$1.71291900 \cdot 10^{-7}$
15	CH_4O	$2.39078180 \cdot 10^{-7}$	$3.37649200 \cdot 10^{-8}$
16	CH_3	$8.36947330 \cdot 10^{-8}$	$1.03174400 \cdot 10^{-8}$
17	H	$4.11784690 \cdot 10^{-8}$	$2.30140300 \cdot 10^{-9}$
18	C_2H_2	$2.73892280 \cdot 10^{-8}$	$3.66416600 \cdot 10^{-9}$
19	Cl	$9.54162630 \cdot 10^{-9}$	$6.57052700 \cdot 10^{-9}$
20	HCLCO	$8.18813620 \cdot 10^{-9}$	$1.09139070 \cdot 10^{-9}$
21	COH	$4.44684800 \cdot 10^{-9}$	$5.67195800 \cdot 10^{-10}$
22	C_2H_3Cl	$9.68524750 \cdot 10^{-10}$	$1.85486650 \cdot 10^{-10}$
23	C_2H_5	$5.38361690 \cdot 10^{-10}$	$1.16769830 \cdot 10^{-10}$
24	CO_2H	$4.12310050 \cdot 10^{-10}$	$6.20315000 \cdot 10^{-11}$
25	C_2H_6O	$3.58515800 \cdot 10^{-10}$	$8.79820100 \cdot 10^{-11}$
26	OH	$2.57008980 \cdot 10^{-10}$	$2.94974900 \cdot 10^{-11}$
27	NH_2	$2.17108990 \cdot 10^{-10}$	$3.14748900 \cdot 10^{-11}$
28	CLCN	$1.33412850 \cdot 10^{-10}$	$1.97532400 \cdot 10^{-11}$
29	Cl_2	$7.74727370 \cdot 10^{-11}$	$9.19055700 \cdot 10^{-12}$
30	CH_2Cl_2	$4.92446890 \cdot 10^{-11}$	$8.93137500 \cdot 10^{-12}$
31	CH_2Cl	$4.60185580 \cdot 10^{-11}$	$8.24734600 \cdot 10^{-12}$
32	C_2H_3	$3.29258890 \cdot 10^{-11}$	$6.66014200 \cdot 10^{-12}$
33	COH_3	$1.65048310 \cdot 10^{-11}$	$3.27816200 \cdot 10^{-12}$
34	ClCO	$1.64072650 \cdot 10^{-11}$	$2.74745400 \cdot 10^{-12}$
35	C_3O_2	$9.89195550 \cdot 10^{-12}$	$1.91459550 \cdot 10^{-12}$
36	HNC	$6.10235940 \cdot 10^{-12}$	$1.04886310 \cdot 10^{-12}$
37	Cl_2CO	$4.10790650 \cdot 10^{-12}$	$7.23910200 \cdot 10^{-13}$
38	N_2C	$2.91385140 \cdot 10^{-12}$	$5.38667100 \cdot 10^{-13}$
39	C_2N_2	$2.27546370 \cdot 10^{-12}$	$4.62061200 \cdot 10^{-13}$
40	C_2HCl	$1.65256150 \cdot 10^{-12}$	$3.30202500 \cdot 10^{-13}$

TABLE 4.1 *(Continued)*

No. p/p	Chemical compound	α_i	$d\alpha_i/dp$, $\dfrac{1}{MPa}$
41	NO	$1.25348100 \cdot 10^{-12}$	$1.92042300 \cdot 10^{-13}$
42	HOCl	$5.94338240 \cdot 10^{-13}$	$1.01609590 \cdot 10^{-14}$
43	CH_3O	$1.62497130 \cdot 10^{-13}$	$3.50140500 \cdot 10^{-14}$
44	CH	~ 0	~ 0
45	CH_2	~ 0	~ 0
46	O	~ 0	~ 0
47	O_2	~ 0	~ 0
49	HO_2	~ 0	~ 0
50	N_2O	~ 0	~ 0
51	N	~ 0	~ 0
52	NH	~ 0	~ 0
53	H_2O_2	~ 0	~ 0

TABLE 4.2 Constants of Direct and Reverse Chemical Reactions.

No. p/p	Chemical reaction	k'	n	C	k''	m	D
1	$OH + H_2 = H_2O + H$	$2.24 \cdot 10^{13}$	0.00	2637	$1.27 \cdot 10^{14}$	-0.03	10280
2	$H + OH = H_2O$	$2.00 \cdot 10^{22}$	-2.00	0	$2.82 \cdot 10^{23}$	-1.98	60090
3	$OH + OH = H_2O + O$	$6.34 \cdot 10^{12}$	0.00	503	$6.84 \cdot 10^{13}$	-0.01	9171
4	$OH + H_2O_2 = H_2O + HO_2$	$1.00 \cdot 10^{13}$	0.00	960	$1.03 \cdot 10^{12}$	0.38	18130
5	$H + H_2O_2 = H_2O + OH$	$7.05 \cdot 10^{12}$	0.00	2100	$1.39 \cdot 10^{09}$	0.87	36330
6	$H + HO_2 = H_2O + O$	$3.30 \cdot 10^{12}$	0.00	503	$7.00 \cdot 10^{10}$	0.45	28350
7	$OH + HO_2 = H_2O + O_2$	$5.01 \cdot 10^{13}$	0.00	503	$1.44 \cdot 10^{14}$	0.17	36890
8	$O + H_2O_2 = H_2O + O_2$	$1.40 \cdot 10^{13}$	0.00	3221	$3.65 \cdot 10^{11}$	0.59	45930
9	$CH + CHO = CH_2 + CO$	$3.16 \cdot 10^{10}$	0.70	503	$6.19 \cdot 10^{10}$	0.89	43320
10	$CH + CO_2 = CHO + CO$	$1.00 \cdot 10^{10}$	0.50	3020	$1.03 \cdot 10^{07}$	1.08	35920
11	$CH + O = CO + H$	$5.01 \cdot 10^{11}$	0.50	0	$1.70 \cdot 10^{13}$	0.54	88510
12	$CHO + CHO = CH_2O + CO$	$1.58 \cdot 10^{11}$	0.50	0	$4.54 \cdot 10^{12}$	0.64	35410
13	$CHO + CH_2 = CH_2 + CO$	$3.16 \cdot 10^{10}$	0.70	503	$4.14 \cdot 10^{11}$	0.81	47280
15	$CHO + H = CO + H_2$	$1.58 \cdot 10^{12}$	0.50	0	$4.18 \cdot 10^{11}$	0.89	44050
16	$CHO + OH = CO + H_2O$	$1.00 \cdot 10^{14}$	0.00	0	$1.50 \cdot 10^{14}$	0.35	51690
17	$CHO + O = CO + OH$	$1.00 \cdot 10^{14}$	0.00	0	$1.38 \cdot 10^{13}$	0.37	43020

18	$CHO + O_2 = CO + HO_2$	$3.16 \cdot 10^{12}$	0.00	3523	$1.66 \cdot 10^{12}$	0.18	18830
19	$CO + OH = CO_2 + H$	$1.26 \cdot 10^{07}$	1.30	-403	$2.99 \cdot 10^{12}$	0.39	12190
20	$CO + HO_2 = CO_2 + OH$	$1.00 \cdot 10^{14}$	0.00	11575	$4.62 \cdot 10^{16}$	-0.44	43350
21	$CO + O = CO_2$	$6.31 \cdot 10^{15}$	0.00	2063	$1.95 \cdot 10^{21}$	-0.87	66080
22	$CO_2 + O = CO + O_2$	$2.51 \cdot 10^{12}$	0.00	22043	$1.44 \cdot 10^{09}$	0.62	17980
23	$CH + O_2 = CO + OH$	$1.26 \cdot 10^{11}$	0.67	12934	$3.15 \cdot 10^{10}$	0.99	92920
24	$CH_2 + HO_2 = CH_2 + O_2$	$1.00 \cdot 10^{11}$	0.50	3020	$3.08 \cdot 10^{12}$	0.60	32070
25	$CO + N_2O = CO_2 + N_2$	$1.00 \cdot 10^{11}$	0.00	10065	$2.67 \cdot 10^{15}$	0.08	53970
26	$N + NH = N_2 + H$	$6.31 \cdot 10^{11}$	0.50	0	$3.56 \cdot 10^{13}$	0.41	75930
27	$H + HO_2 = H_2 + O_2$	$2.51 \cdot 10^{13}$	0.00	352	$1.26 \cdot 10^{13}$	0.20	29090
28	$N_2 = N + N$	$3.98 \cdot 10^{21}$	-1.60	113236	$6.84 \cdot 10^{19}$	-1.49	-468
29	$CH_2 + NH = CH_2 + N$	$1.00 \cdot 10^{13}$	0.00	2516	$1.47 \cdot 10^{15}$	-0.02	17500
30	$OH + Cl = HCl + O$	$7.24 \cdot 10^{12}$	0.00	3800	$1.16 \cdot 10^{11}$	0.00	3340
31	$H_2 = H + H$	$2.00 \cdot 10^{14}$	0.00	48314	$8.05 \cdot 10^{13}$	-0.05	-4127

Taking into account eq 4.9 noted above for substances with numbers 1–7 will correspond in the following form:

$$\frac{d\alpha_1}{dt} = \frac{G_m(\alpha_{10} - \alpha_1)}{\rho W} + K_9' \cdot \alpha_{21}\alpha_{44} - K_9'' \cdot \alpha_1\alpha_{45} + K_{10}' \cdot \alpha_5\alpha_{44} - K_{10}'' \cdot \alpha_1\alpha_{21} +$$

$$+ K_{11}' \cdot \alpha_{44}\alpha_{46} - K_{11}'' \cdot \alpha_1\alpha_{17} + K_{12}' \cdot \alpha_{21}^2 - K_{12}'' \cdot \alpha_1\alpha_{11} + K_{13}' \cdot \alpha_{21}\alpha_{45} - K_{13}'' \cdot \alpha_1\alpha_{16} +$$
$$+ K_{14}' \cdot \alpha_{16}\alpha_{21} - K_{14}'' \cdot \alpha_1\alpha_7 + K_{15}' \cdot \alpha_{17}\alpha_{21} - K_{15}'' \cdot \alpha_1\alpha_2 + K_{16}' \cdot \alpha_{21}\alpha_{27} - K_{16}'' \cdot \alpha_1\alpha_4 +$$
$$+ K_{17}' \cdot \alpha_{21}\alpha_{46} - K_{17}'' \cdot \alpha_1\alpha_{27} + K_{18}' \cdot \alpha_{21}\alpha_{47} - K_{18}'' \cdot \alpha_1\alpha_{48} - K_{19}' \cdot \alpha_1\alpha_{27} + K_{19}'' \cdot \alpha_5\alpha_{17} -$$
$$- K_{20}' \cdot \alpha_1\alpha_{48} + K_{20}'' \cdot \alpha_5\alpha_{27} - K_{21}' \cdot \alpha_1\alpha_{46} + K_{21}'' \cdot \alpha_5 + K_{22}' \cdot \alpha_5\alpha_{46} - K_{22}'' \cdot \alpha_1\alpha_{47} +$$
$$+ K_{23}' \cdot \alpha_{44}\alpha_{47} - K_{23}'' \cdot \alpha_1\alpha_{27} - K_{25}' \cdot \alpha_1\alpha_{49} + K_{25}'' \cdot \alpha_3\alpha_5;$$

$$\frac{d\alpha_2}{dt} = \frac{G_m(\alpha_{20} - \alpha_2)}{\rho W} + K_1' \cdot \alpha_2\alpha_{17} - K_1'' \cdot \alpha_4\alpha_{17} + K_{15}' \cdot \alpha_{17}\alpha_{21} - K_{15}'' \cdot \alpha_1\alpha_2 +$$

$$+ K_{27}' \cdot \alpha_{17}\alpha_{48} - K_{27}'' \cdot \alpha_2\alpha_{47} - K_{31}' \cdot \alpha_2 + K_{31}'' \cdot \alpha_{17}^2;$$

$$\frac{d\alpha_3}{dt} = \frac{G_m(\alpha_{30} - \alpha_3)}{\rho W} + K_{25}' \cdot \alpha_1\alpha_{49} - K_{25}'' \cdot \alpha_3\alpha_5 + K_{26}' \cdot \alpha_{50}\alpha_{51} - K_{26}'' \cdot \alpha_3\alpha_{17} - K_{28}' \cdot \alpha_3 + K_{28}'' \cdot \alpha_{50}^2$$

$$\frac{d\alpha_4}{dt} = \frac{G_m(\alpha_{40} - \alpha_4)}{\rho W} + K_1' \cdot \alpha_2\alpha_{27} - K_1'' \cdot \alpha_4\alpha_{17} + K_2' \cdot \alpha_{17}\alpha_{27} - K_2'' \cdot \alpha_4 +$$

$$+ K_3' \cdot \alpha_{27}^2 - K_3'' \cdot \alpha_4\alpha_{46} + K_4' \cdot \alpha_{27}\alpha_{52} - K_4'' \cdot \alpha_4\alpha_{48} + K_5' \cdot \alpha_{17}\alpha_{52} - K_5'' \cdot \alpha_4\alpha_{27} +$$

$$+ K_6' \cdot \alpha_{17}\alpha_{48} - K_6'' \cdot \alpha_4\alpha_{46} + K_7' \cdot \alpha_{27}\alpha_{48} - K_7'' \cdot \alpha_4\alpha_{47} + K_8' \cdot \alpha_{46}\alpha_{52} - K_8'' \cdot \alpha_4\alpha_{47} +$$

$$+ K_{16}' \cdot \alpha_{21}\alpha_{27} - K_{16}'' \cdot \alpha_1\alpha_4;$$

$$\frac{d\alpha_5}{dt} = \frac{G_m(\alpha_{50} - \alpha_5)}{\rho W} + K_{10}' \cdot \alpha_5\alpha_{44} - K_{10}'' \cdot \alpha_1\alpha_{21} + K_{19}' \cdot \alpha_4\alpha_{27} - K_{19}'' \cdot \alpha_5\alpha_{17} +$$

$$+ K_{20}' \cdot \alpha_1\alpha_{48} - K_{20}'' \cdot \alpha_5\alpha_{27} + K_{21}' \cdot \alpha_1\alpha_{46} - K_{21}'' \cdot \alpha_5 - K_{22}' \cdot \alpha_5\alpha_{46} + K_{22}'' \cdot \alpha_1\alpha_{47} +$$

$$+ K_{25}' \cdot \alpha_1\alpha_{49} - K_{25}'' \cdot \alpha_3\alpha_5;$$

$$\frac{d\alpha_6}{dt} = \frac{G_m(\alpha_{60} - \alpha_6)}{\rho W} + K_{30}' \cdot \alpha_{19}\alpha_{27} - K_{30}'' \cdot \alpha_6\alpha_{46}$$

$$\frac{d\alpha_7}{dt} = \frac{G_m(\alpha_{70} - \alpha_7)}{\rho W} + K_{14}' \cdot \alpha_{16}\alpha_{21} - K_{14}'' \cdot \alpha_1\alpha_7 + K_{24}' \cdot \alpha_{16}\alpha_{48} - K_{24}'' \cdot \alpha_7\alpha_{47} +$$

$$+ K_{29}' \cdot \alpha_{16}\alpha_{51} - K_{29}'' \cdot \alpha_7\alpha_{50}$$

$$a_{11} = -\frac{G_m}{\rho W} - K_9''\alpha_{45} - K_{10}''\alpha_{21} - K_{11}''\alpha_{17} - K_{12}''\alpha_{11} - K_{13}''\alpha_{16} - K_{14}''\alpha_7 - K_{15}''\alpha_2 - K_{16}''\alpha_4 -$$

$$- K_{17}''\alpha_{27} - K_{18}''\alpha_{48} - K_{19}'\alpha_{27} - K_{20}'\alpha_{48} - K_{21}'\alpha_{46} - K_{22}'\alpha_{47} - K_{23}'\alpha_{27} - K_{25}'\alpha_{49};$$

$$a_{12} = -K_{15}''\alpha_1; \quad a_{13} = K_{25}''\alpha_5; \quad a_{14} = -K_{16}''\alpha_1;$$

$$a_{15} = K_{10}'\alpha_{44} + K_{19}''\alpha_{17} + K_{20}''\alpha_{27} + K_{21}'' + K_{22}'\alpha_{46} + K_{25}''\alpha_3; \quad a_{16} = 0; \quad a_{17} = -K_{14}''\alpha_1;$$

$$a_{21} = -K_{15}''\alpha_2; \quad a_{22} = -\frac{G_m}{\rho W} - K_1'\alpha_{17} - K_{15}''\alpha_1 - K_{27}''\alpha_{47} - K_{31}'\alpha_{45};$$

$$a_{23} = a_{24} = a_{25} = a_{26} = a_{27} = 0;$$

$$a_{31} = K_{25}'\alpha_{49}; \quad a_{32} = 0; \quad a_{33} = -\frac{G_o}{\rho W} - K_{25}''\alpha_5 - K_{26}''\alpha_{17} - K_{28}'\alpha_{18};$$

$$a_{34} = 0; \quad a_{35} = -K_{25}''\alpha_3; \quad a_{36} = a_{37} = 0;$$

$$a_{41} = -K_{16}''\alpha_4; \quad a_{42} = K_1'\alpha_{27}; \quad a_{43} = 0;$$

$$a_{44} = -\frac{G_m}{\rho W} - K_1''\alpha_{17} - K_2''\alpha_{25} - K_3''\alpha_{46} - K_4''\alpha_{48} - K_5''\alpha_{27} - K_6''\alpha_{46} - K_7''\alpha_{47} - K_8''\alpha_{47} - K_{16}''\alpha_1;$$

$$a_{45} = a_{46} = a_{47} = 0;$$

$$a_{51} = K_{10}'\alpha_{21} + K_{20}''\alpha_{48} + K_{21}''\alpha_{46} + K_{22}''\alpha_{47} + K_{25}'\alpha_{49}; \quad a_{52} = 0;$$

$$a_{53} = -K_{25}''\alpha_5; \quad a_{54} = K_{19}'\alpha_{27};$$

$$a_{55} = -\frac{G_m}{\rho W} - K_{10}'\alpha_{44} - K_{19}''\alpha_{17} - K_{20}''\alpha_{27} - K_{21}''\alpha_{24} - K_{22}'\alpha_{46} - K_{25}''\alpha_3; \quad a_{56} = a_{57} = 0;$$

$$a_{61} = a_{62} = a_{63} = a_{64} = a_{65} = 0; \quad a_{66} = -\frac{G_m}{\rho W} - K''_{30}\alpha_{46}; \quad a_{67} = 0;$$

$$a_{71} = -K''_{14}\alpha_7; \quad a_{72} = a_{73} = a_{74} = a_{75} = a_{76} = 0;$$

$$a_{77} = -\frac{G_m}{\rho W} - K''_{14}\alpha_1 - K''_{24}\alpha_{47} - K''_{29}\alpha_{50}$$

Based on the written relations, the Jacobian (4.10) will look like this:

$$\mathbf{J} = \begin{bmatrix} -39.64 \cdot 10^{12} & -99.48 \cdot 10^{9} & -52.44 \cdot 10^{13} & -35.72 \cdot 10^{12} & -66.75 \cdot 10^{12} & 0 & -12.09 \cdot 10^{11} \\ -5,38 \cdot 10^{10} & 19,21 \cdot 10^{13} & 0 & 0 & 0 & 0 & 0 \\ 0 & 0 & -5.23 \cdot 10^{14} & 0 & -6.67 \cdot 10^{13} & 0 & 0 \\ -3.93 \cdot 10^{13} & 6.12 \cdot 10^{3} & 0 & -35.7 \cdot 10^{12} & 0 & 0 & 0 \\ 50.23 & 0 & -5.23 \cdot 10^{14} & 0.0032 & -6.67 \cdot 10^{13} & 0 & 0 \\ 0 & 0 & 0 & 0 & 0 & 0 & 0 \\ -33.73 \cdot 10^{10} & 0 & 0 & 0 & 0 & 0 & 12.13 \cdot 10^{11} \end{bmatrix}$$

The vector (4.11) contains the eigenvalues of the Jacobian (4.10)

$$A = \begin{Bmatrix} 1.954 \cdot 10^{14} \\ -7.495 \cdot 10^{13} \\ 1.343 \cdot 10^{12} \\ -1.936 \cdot 10^{11} \\ -5.621 \cdot 10^{14} \\ -356.171 \\ 0 \end{Bmatrix} \tag{4.11}$$

As in a vector (4.11), there are no complex numbers, according to the researches provided in Ref. [12] it is possible to speak about absence of influence of chemical responses to pressure fluctuations in the combustion chamber.

4.3   CONCLUSION

Thus, in the study of the eigenvalues of the matrix of the linearized model, the absence of influence of chemical reactions on pressure fluctuations in the combustion chamber was revealed.

In conclusion, we note that the conclusions obtained cannot be generalized to various types of fuels for thermal engines, since the availability of data on chemical reactions is very limited. Nevertheless, the application of the developed mathematical models seems justified for all types of propulsion systems.

KEYWORDS

- **rocket engine**
- **propellant**
- **chemical equilibrium**
- **internal ballistics**
- **eigenvalues**
- **linearization**
- **frequency**

REFERENCES

1. Aliev A.V. Internal ballistics / RARAN; Under the editorship of Lipanov A.M., MilekhinYu.M. – M.: Mechanical engineering, 2007. – 504 p. (in Russ).
2. Solomonov, Yu. S.; Lipanov, A. M.; Aliev, A. V.; Dorofeev, A. A.; Cherepov, V. I. *Solid Propellant Regulable Rocket Engine/RARAN*. Mechanical Engineering, 2011, p 416 (in Russian).
3. Assovskiy I.G. Physics of burning and internal ballistics. M.: Science, 2005. – 357 p. (in Russ).
4. L.K. Gusachenko, V.E. Zarko, V.I.Ziriynov, V.P.Bovrishev. Simulation of processes of burning of solid fuels/ Novosibirsk: Science, 1985. – 182 p. (in Russ).
5. Lipanov A.M. Physical and chemical and mathematical models of burning of compound solid missile fuels: preprint. – Izhevsk: Institute computer researches, 2007. – 112 p. (in Russ).
6. Varnatts Yu., Maas At., Dibbl R. Burning physical and chemical aspects, modeling, experiments, formation of the polluting substances / Lane from English Agafonov G.

L. Under the editorship of Vlasov P. A. – M.: Publishing house Fizmatlit, 2003. – 352 p. (in Russ).

7. Raushenbakh, B. V. *Vibrational Burning*; M.: prod.: Fizmatlit, 1961 (in Russian).

8. Research of Missile Engines on Solid Fuel; Under M. Sammerfild's edition. *Foreign Literature*, 1963; p 440 (in Russian).

9. Williams F. A. Theory of burning. – M.: Science. -1971. – 616 p. (in Russ).

10. Salnikov A.F., Petrova E.N. The pilot study of influence of high-frequency oscillations on longitudinal acoustic instability of the missile engine on solid fuel//the Bulletin of MSTU of N.E. Bauman. Mechanical engineering series. -2009. – N9. Page 89–97. (in Russ).

11. Salnikov, A. F.; Petrova, E. N.; Baluyeva, M. A. Influence of Structural Elements of the Combustion Chamber of the Solid-fuel Missile Engine on Value of a Vibration Amplitude of Pressure. Bulletin of the Perm National Research Polytechnical University. *Space Tech.* 2006, *26*, N. 26, p 16–20 (in Russian).

12. Davydov, Yu. M.; Davydova, I. M.; Egorov, M. Yu. Instability of Worker Process in the Two-chamber Missile Engine on Solid Fuel. Reports of Academy of Sciences, 2011, T. 439, N.2, pp 188–191 (in Russian).

13. Larionov V. M., Zaripov R. G. Auto-oscillations of gas in installations with burning. - Kazan: prod. Kazan state technical university, 2003. – 227 p. (in Russ).

14. Sabdenov, K. O. The Modes of Burning of the Solid Missile Fuel Which Is Breaking up to Gas on the Pyrolysis Mechanism. News of the Tomsk Polytechnic University, 2006, T. 309, N. 3, p 120–125 (in Russian).

15. Myrzakulov, R.; Kozybakov, M. Zh.; Sabdenov, K. O. Extinction of Burning of Solid Missile Fuels and Explosives in Case of Live Pressure. News of the Tomsk Polytechnic University, 2006, T. 309, N. 5, p 123–130.

16. Kudryavtsev I. K. Chemical instability. – M.: prod. Moscow university, 1987. – 254 p. (in Russ).

17. Aliev, A. V.; Mishchenkova, O. V. About Application of a Method of Linearization in Case of the Decision of Some Tasks of Internal Ballistics. The Bulletin of the Pacific State Technical University, 2007, N. 4 (7), p 25–38 (in Russian).

18. Aliev A. V., Voevodina O. A., Pushina E. S. Page of Model of nonstationary thermogasdynamic processes in missile engines taking into account chemical equilibrium of combustion products//Science and education, Al. No. FS 77-48211, 2015. - No. 11. Page 253–256. (in Russ).

19. Aliev, A. V.; Mishchenkova, O. V. Mathematical Modeling in Equipment. Institute of Computer Researches: M – Izhevsk, 2012; p 476 (in Russian).

20. Aliev, A. V.; Mishchenkova, O. V. *Mathematical Modeling and Numerical Methods in Chemical Physics and Mechanics*. Apple Academic Press: Waretown, Oakville, 2016; p 544.

CHAPTER 5

CORRELATION AND JUSTIFICATION OF THE FORMATION OF SOME IONIC SOLIDS IN NATURE FROM THE NATURAL SOURCES OF THE ATOMS AND IONS AND IN TERMS OF THEIR COMPUTED LATTICE ENERGIES

SARMISTHA BASU[1] and NAZMUL ISLAM[2,*]

[1]*Department of Electronics, Behala College, Kolkata 700060, West Bengal, India*

[2]*Theoretical and Computational Research Laboratory, Ramgarh Engineering College, Jharkhand 825101, India*

Corresponding author. E-mail: nazmul.islam786@gmail.com

ABSTRACT

It is found that nature picks up energetically the most favorable route in the formation of an ionic crystal from their constituents.

We see in nature that all solid bodies have beautiful crystalline structures—which must be the most stable state of occurrence of the constituents of the crystals. In this work, we have made a quest of all energetic effects and the reasons for formation of some ionic solids from their constituents at their primitive states of occurrence in nature. It is found that nature picks up energetically the most favorable route in the formation of an ionic crystal from their constituents in the primitive state. In this study, we have relied upon a correlation involving heat of sublimation, enthalpy of formation, lattice energy, and a very important physical data—the melting point of the ionic solids. We have worked on 15 ionic solids in this study. Our investigation reveals that in the formation of ionic solids in nature, the laws of thermodynamics prevail.

5.1   INTRODUCTION

If we look at nature, we find that the majority of solids have some unique ingredients—crystals—both ionic and covalent.

It is said that diamond is a very hard solid, but very slowly it is converted to graphite, but the reverse never occurs.

In this study, we are in search of the rationale of formation of crystals of some ionic compounds.

It is well known that the experimental determination of enthalpy of formation of solids is a very difficult and involved process. Similarly, the experimental determination of the lattice energies of crystalline compounds is also extremely difficult and the result is uncertain and erroneous.

However, scientists like Donald, Jenkins, and Leslie Glasser,[1] Kapustinkii,[2] Yatsimirskii,[3,4] Waddington,[5] Gale,[6] Busing,[7] Catlow,[8] Jenkins and Roobottom,[9,10] Morell,[11] and Ghosh and Biswas[12] suggested an empirical formula for the theoretical determination of the lattice energies of the ionic solids. Of these, the Born–Lande equation has the clear visualization of the attraction and the repulsion of the infinite number of ions including the repulsive forces/energies between them due to the overlapping of the charge clouds of the ions and can compute the lattice energy in a simpler way. This equation is the most popular one among the scientists involved in the research to find out the rationale of energetics of the formation of ionic compounds.[13–18]

Ionic solids are solids composed of oppositely charged ions. In most ionic compounds, the anions are much larger than the cations, and it is the anions which form the crystal array.[19–22] In an ionic compound, cations and anions are arranged in space to form an extended 3-D array that maximizes the number of attractive electrostatic interactions and minimizes the number of repulsive electrostatic interactions. These types of solids are generally high melting, hard and brittle, water soluble, and dissolve to form solutions that are electrical conductors (Fig. 5.1).

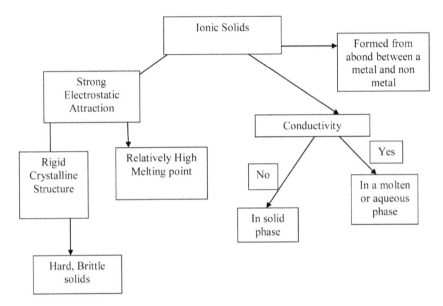

FIGURE 5.1   Bonding classifications of ionic solids.

The smaller cations reside in the holes between the anions.

(1) Ions are assumed to be charged, incompressible, and non-polarizable spheres.

(2) Ions try to surround themselves with as many ions of opposite charge as closely as possible. Usually, in the packing arrangement, the cation is just large enough to allow the anions to surround it without touching one another.

(3) The cation to anion ratio must reflect the stoichiometry of the compound.

Lattice energy is a type of potential energy. The lattice energy is the energy required to break apart an ionic solid and convert its component atoms into gaseous ions. It causes the value for the lattice energy to always be positive since this will always be an endothermic reaction. The lattice energy is the reverse process, meaning it is the energy released when gaseous ions bind to form an ionic solid. This process will always be exothermic, and thus the value for lattice energy will be negative. Lattice energy is used to explain the stability of ionic solids. Some might expect such an ordered structure to be less stable because the entropy of the system would be low.[23–30]

However, the crystalline structure allows each ion to interact with multiple oppositely charged ions, which causes a highly favorable change in the enthalpy of the system. A lot of energy is released as the oppositely charged ions interact. It is this that causes ionic solids to have such high melting and boiling points. Some require such high temperatures that they decompose before they can reach a melting and/or boiling point.[31–40]

There are several important concepts to understand before the Born–Haber cycle can be applied to determine the lattice energy of an ionic solid: ionization energy, electron affinity, dissociation energy, sublimation energy, and heat of formation.

Ionization energy is the energy required to remove an electron from a neutral atom or an ion. This process always requires an input of energy, and thus will always have a positive value. Ionization energy increases across the periodic table from left to right and decreases from top to bottom. There are some exceptions, usually due to the stability of half-filled and completely filled orbitals. Ionization energy is the amount of energy needed by a gaseous atom in order to remove an electron from its outermost orbital. It describes absorption energy from outside. It is used to describe an electron moving. The ionization energy of an element increases as one moves across a period in the periodic table because the electrons are held tighter by the higher effective nuclear charge. This energy of the elements increases as one moves up a given group because the electrons are held in lower energy orbitals, closer to the nucleus and therefore are more tightly bound (harder to remove).[41–47]

The ionization energy of a chemical species (i.e., an atom or molecule) is the energy required to remove electrons from gaseous atoms or ions. This property is also referred to as the ionization potential and is measured in volts. In chemistry, it often refers to one mole of a substance (molar ionization energy or enthalpy) and is reported in kJ/mol. In atomic physics, the ionization energy is typically measured in the unit electron volt (eV). Large atoms or molecules have low ionization energy, while small molecules tend to have higher ionization energies.[48–53]

The ionization energy is different for electrons of different atomic or molecular orbitals. More generally, the nth ionization energy is the energy required to strip off the nth electron after the first $n - 1$ electrons have been removed. It is considered a measure of the tendency of an atom or ion to surrender an electron or the strength of the electron binding. The greater the ionization energy, the more difficult it is to remove an electron.

The ionization energy may be an indicator of the reactivity of an element. Elements with a low ionization energy tend to be reducing agents and form cations, which in turn combine with anions to form salts. The ionization energy of an element increases as one moves across a period in the periodic table because the electrons are held tighter by the higher effective nuclear charge. This is because additional electrons in the same shell do not substantially contribute to shielding each other from the nucleus; however, an increase in atomic number corresponds to an increase in the number of protons in the nucleus.

The ionization energy of the elements increases as one moves up a given group because the electrons are held in lower energy orbitals, closer to the nucleus and thus more tightly bound (harder to remove). Based on these two principles, the easiest element to ionize is francium and the hardest to ionize is helium.

Electron affinity is the energy released when an electron is added to a neutral atom or an ion. The energy released would have a negative value, but due to the definition of electron affinity, it is written as a positive value in most tables. Therefore, when used in calculating the lattice energy, we must remember to subtract the electron affinity, not add it. Electron affinity increases from left to right across the periodic table and decreases from top to bottom. The electron affinity (Eea) of an atom or molecule is considered to be the amount of energy released or spent when an electron is added to a neutral atom or molecule in the gaseous state to form a negative ion:

$$X + e^- \rightarrow X^- + \text{energy.}$$

Electron affinity describes the release of energy to a surrounding. It is the energy change for a process where an atom gains an electron. It used to describe electron gaining. The energy will always be negative. Moving down a group will decrease electron affinity (less negative). This is again because the electrons get far from the nucleus. Moving right will increase the electron affinity (more negative). This is because of increase in change.

Exceptions again include first and fourth electrons in p-block, Period 3 is greater than Period 2 (Fig. 5.2).

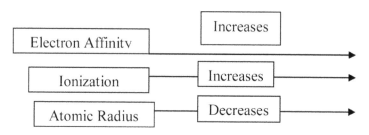

FIGURE 5.2 Periodic table of the elements.

Dissociation energy is the energy required to break apart a compound. The dissociation of a compound is always an endothermic process, meaning it will always require an input of energy. Therefore, the change in energy is always positive. The magnitude of the dissociation energy depends on the electronegativity of the atoms involved.

Sublimation energy is the energy required to cause a change of phase from solid to gas, bypassing the liquid phase. This is an input of energy and thus has a positive value. It may also be referred to as the energy of atomization.

The heat of formation is the change in energy when forming a compound from its elements. This may be positive or negative, depending on the atoms involved and how they interact.

5.2 BORN–HABER CYCLE

The formation of crystals of ionic compounds is visualized in a pictorial cycle called Born–Haber cycle—where the steps involved in the formation of the ionic solids from the initial states of occurrence of the elements involved, are beautifully depicted.

We now depict pictorially the steps involving the Born–Haber cycle in Figure 5.3.

Born - Haber Cycle

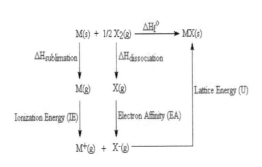

$$\Delta H_f^0 = \Delta H_{sub} + IE + \Delta H_{diss} + EA + U$$

FIGURE 5.3 Increment and decrement phenomenon between electron affinity, ionization, and atomic radius.

The Born–Haber cycle can be reduced to a single equation as follows:

Heat of formation = Heat of atomization + Dissociation energy + Ionization energies + Sum of electron affinities + Lattice energy.

In this general equation, the electron affinity is added. However, when plugging in a value, determine whether energy is released (exothermic reaction) or absorbed (endothermic reaction) for each electron affinity. If energy is released, put a negative sign in front of the value; if energy is absorbed, the value should be positive.

Rearrangement to solve for lattice energy gives the following equation:

Lattice energy = Heat of formation − Heat of atomization − Dissociation energy − Sum of ionization energies − Sum of electron affinities.

The Born–Lande equation of lattice energy is as follows:

$$U = \frac{N_A M z^+ z^- e^2}{4\pi\varepsilon_0 r_0}\left(1 - \frac{1}{n}\right),$$

where,

N_A is the Avogadro constant;

M is the Madelung constant, relating to the geometry of the crystal;

z^+ is the charge number of cation;

z^- is the charge number of anion;

q_e is the elementary charge, equal to 1.6022×10^{-19} C;

ε_0 is the permittivity of free space, equal to 8.854×10^{-12} C^2 J^{-1} m^{-1};

r_0 is the distance between closest ions and

n is the Born exponent, a number between 5 and 12, determined experimentally by measuring a compressibility of the solid, or derived theoretically.

The necessary steps of formation of ionic solids are considered from the natural occurrence of interacting ionic species. In this thermos-chemical cycle, the first input is the sublimation of atomic solids, the second step is the ionization of metal atoms to metal ions. Regarding the non-metal part of ionic compounds, usually the molecules occur in the gaseous state and the bond between the atoms has to be broken by supplying energy—the "Bond Dissociation Energy."

Next step is to convert the atoms of non-metals in the ionic state—anions by adding electrons to such atoms. The energetic effect in this process is well known as electron affinity. In majority of cases, the electron affinity is negative. Literature shows that the lattice energy, a type of potential energy that relates to the stability of ionic solids, is computed theoretically in terms of some suggested formula. The balance sheet shows all the energetic effects—namely the heat of sublimation, the ionization energy, the bond dissociation energy, the electron affinity, and the lattice energy.

From the balance sheet of these energetic effects, one can get the enthalpy of formation of the ionic compounds. We have taken up the present work to find a detailed rationale of formation of a number of ionic crystals from the initial states of occurrence of the elements of the crystal. If we allow the formation of sodium chloride crystal from the elements of sodium and chlorine in their natural states of occurrence, the steps are as follows. In the above, we have noted that, prior to the formation of ionic crystal, we have to generate one mole of sodium ions from one mole of sodium in its natural occurrence. The steps required are sublimation from natural occurrence of sodium atoms and ionization of sodium atoms.

Similarly, the steps of generation of chloride ions are—first molecules are dissociated by the supply of bond dissociation energy and then putting the electrons on the atoms and the atoms are converted into ions and the energetic effect is the electron affinity.

Let us investigate the energetic effects in the formation of various ensembles of Avogadro number sodium ions and Avogadro number of chloride ions.

The ensemble may occur as one ion pair, ion square, and ion cube. Let us try the energetic effects where the ion pair, ion square, and ion cube which are all thermodynamically allowed ensemble of sodium and chloride ions. But it is the ionic lattice and neither the ion pairs nor the ion squares or ion cubes, the other energetically allowed alignments of cations and anions—that are formed during the process of formation sodium chloride from its natural sources (Tables 5.1–5.4).

TABLE 5.1 Computation of the Heat of Formation of Solid Ionic Compounds.

Compounds	Sublimation enthalpy (kJ/mol)	Bond dissociation energy (kJ/mol)	Ionization energy (kJ/mol)	Electron affinities (kJ/mol)	Lattice energy (kJ/mol)	Born–Haber cycle (ΔH_f) = ΔH_s + 1/2 BDE + IE + EA + LE
LiF	155.2	158	520	−328	−1036	−616.0
LiCl	155.2	243	520	−349	−853	−408.6
LiBr	155.2	192	520	−342	−807	−351.2
LiI	155.2	151	520	−314	−757	−270.4
NaF	108	158	496	−328	−923	−569.0
NaCl	108	243	496	−349	−786	−411.12
NaBr	108	192	496	−342	−747	−361.1
NaI	108	151	496	−314	−704	−287.8
KF	89	158	419	−328	−821	−562.6
KCl	89	243	419	−349	−715	−436.68
KBr	89	192	419	−342	−682	−392.2
KI	89	151	419	−314	−649	−327.9
MgO	146	498	738	−141	−3800	−601.6
CaO	178	498	590	−141	−3400	−635.09
SrO	164	498	550	141	−3200	−592.0
BaO	180	498	503	−141	−3000	−548.1

TABLE 5.2   The Lattice Energy of Ionic Solids and Their Melting Points.

Compounds	Lattice energy (kJ/mole)	Melting points (°C)
LiF	−1036	848.2
LiCl	−853	610
LiBr	−807	552
LiI	−757	469
NaF	−923	996
NaCl	−786	800.7
NaBr	−747	747
NaI	−704	660
KF	−821	858
KCl	−715	771
KBr	−682	734
KI	−649	681
MgO	−3800	2825
CaO	−3400	2898
SrO	−3200	2531
BaO	−3000	1972

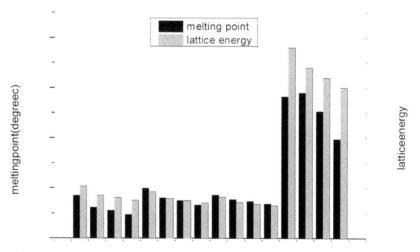

FIGURE 5.4   Correlation plot of lattice energy of ionic solids and their melting points.

TABLE 5.3   The Heat of Formation of Ionic Solids and Their Melting Points.

Compounds	Heat of formation (kJ/mole)	Melting points (°C)
LiF	−616.0	848.2
LiCl	−408.6	610
LiBr	−351.2	552
LiI	−270.4	469
NaF	−569.0	996
NaCl	−411.12	800.7
NaBr	−361.1	747
NaI	−287.8	660
KF	−562.6	858
KCl	−436.68	771
KBr	−392.2	734
KI	−327.9	681
MgO	−601.6	2825
CaO	−635.09	2898
SrO	−592.0	2531
BaO	−548.1	1972

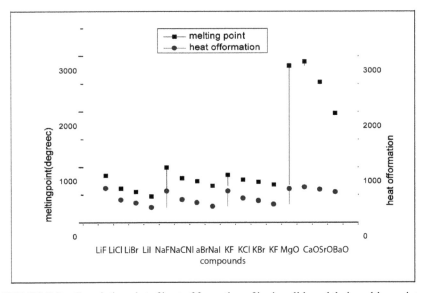

FIGURE 5.5   Correlation plot of heat of formation of ionic solids and their melting points.

TABLE 5.4   The Heat of Formation of Ionic Solids and Lattice Energy.

Compounds	Heat of formation (kJ/mole)	Lattice energy (kJ/mole)
LiF	−616.0	−1036
LiCl	−408.6	−853
LiBr	−351.2	−807
LiI	−270.4	−757
NaF	−569.0	−923
NaCl	−411.12	−786
NaBr	−361.1	−747
NaI	−287.8	−704
KF	−562.6	−821
KCl	−436.68	−715
KBr	−392.2	−682
KI	−327.9	−649
MgO	−601.6	−3800
CaO	−635.09	−3400
SrO	−592.0	−3200
BaO	−548.1	−3000

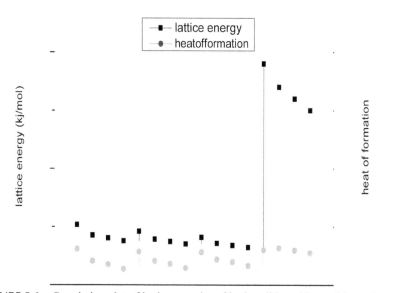

FIGURE 5.6   Correlation plot of lattice energies of ionic solids and heat of formation.

In determining the lattice energy, we use the crystal radii or Shanon radii of the ions. In the lattice, the ions execute harmonic oscillation about the new position of the rest. Therefore, the ions do not remain at rest in contact with each other. Therefore, there is some space in the solids between the cations and anions. Scientists have measured the closest distance between the ions in the solid state and have apportioned this distance between the ions—cation and anion.

We are habituated to at legend to such size data—ionic radii. However, such radii cannot present the absolute size of the ions and must be greater than the absolute size of the radii. Literature shows that Ghosh and Biswas first time computed the absolute radii theoretically.

TABLE 5.5 Shanon's Crystal Radii and the Absolute Radii of Ghosh and Biswas.

Ions	Shanon's radii (A)	Ghosh and Biswas absolute radii
Na^{+1}	1.134	0.3090044
K^{+1}	1.5	0.61452
Li^{+1}	0.757	0.1959889
Ca^{+2}	1.155	0.5442891
Ba^{+2}	1.474	1.0325268
Mg^{+2}	0.71	0.2696408
Cl^{-1}	1.81	0.8282661
F^{-1}	1.306	0.4364289
Br^{-1}	1.96	1.0802367
I^{-1}	2.2	1.4597793
O^{-2}	1.382	0.549787

We cite the Shanon's crystal radii and the absolute radii of Ghosh and Biswas in Table 5.5. From the table, we see that the absolute radii of Ghosh and Biswas is always smaller than the crystal radii of Shanon. This justifies the view that ions do not touch each other in the solid state because they execute harmonic oscillation about the mean position of rest.

5.3 RESULTS AND DISCUSSION

For a clearer visualization of the situation, we have correlated melting points and the lattice energy of the ionic solids in the Figure 5.4. This type of correlation between such quantities is also revealed in a more sophisticated calculation by Laslie Glasser, Donald, and Jenkins.[2]

From Figure 5.5, it is clear that there is a beautiful correlation between the melting point and the heat of formation of ionic compounds. From Figure 5.6, we get the distinct correlation between the heat of formation of the ionic solids and the crystal lattice energy.

5.4 CONCLUSION

During the formation of some ionic solids are beautifully correlated and participated in the formation of ionic solids in the nature and laws of thermodynamics are beautifully reflected in the phenomenon of formation of ionic crystal in nature.

ACKNOWLEDGMENTS

Authors are thankful to Late Prof. Dr. D. C. Ghosh for his valuable suggestion and opinion on the chapter.

Dr. Nazmul Islam is thankful to Techno India Group, Ramgarh Engineering College and Teqip III for necessary help and support.

KEYWORDS

- ionic lattice
- lattice energy
- Born–Haber cycle
- Born–Lande equation

REFERENCES

1. *J. Am. Chem. Soc.* **2000,** *22,* 632–638.
2. Kapustinskii, A. F. Q. *Rev. Chem. Soc.* **1956,** *10,* 283.
3. Yatsimirskii, K. B. *Zh. Obshch. Khim* **1956,** *26,* 94.
4. Yatsimirskii, K. B. *Zh. Neorg. Khim* **1961,** *6,* 518.
5. Waddington, T. C. *Adv. Inorg. Chem. Radiochem.* **1959,** *1,* 157.
6. Gale, J. D. *J. Chem. Soc. Faraday Trans.* **1997,** *93,* 629.

7. W. R. Busing, WMIN, ORNL-5747, Oak Ridge National Laboratory revised version, Mar. 1984. J. D. Gale, GULP—General Utility Lattice Program.

8. C. R. A. Catlow and W. C. Mackrodt. *Computer Simulation of Solids, Lecture Notes in Physics 166*; Springer-Verlag: Berlin, 1982.

9. David M. Birney. Department of Chemistry and Biochemistry, Texas Tech University, Lubbock, TX 79409-1061. *Chem. Educ.* **1999,** *76*, 1570.

10. Lewis, G. N.; Randall, M. J. Am. Chem. Soc. 1921, *CRCnetBASE 1999*, 80th ed.; Lide, D. R., Ed.; Times Mirror Publications: New York, 1999.

11. Martin Senn Andrew Sutherland. The Journal of Organic Chemistry 2019 84 (1), 346–364. *J. Phys. Chem. A* **2005,** *109* (1), 205–212.

12. Int. J. Mol. Sci. 2003, 4, 1–12 (PDF format 250 K). Barbara Graham-Evans1, Paul ..... Issue 6 (June 2003) pages 312–407...2003, 4, 379–407.

13. Heinz, H.; Suter, U. W. *J. Phys. Chem. B* **2004,** *108*, 18341–18352.

14. Electrical Conductivity of Ionic Compound. *Archived from the original on 21 May 2014,* Retrieved December 2, 2012.

15. Carter, Robert. Lattice Energy (PDF). CH370 Lecture Material. Archived (PDF) from the original on May 13, 2015, Retrieved January 19, 2016.

16. Brackett, T. E.; Brackett, E. B. The Lattice Energies of the Alkaline Earth Halides. *J. Phys. Chem.* **1965,** *69* (10), 3611–3614.

17. *Pauling, Linus. The Influence of Relative Ionic Sizes on the Properties of Ionic Compounds. J. Am. Chem. Soc.* **1928.**

18. Sherman, Jack. Crystal Energies of Ionic Compounds and Thermochemical Applications. *Chem. Rev.* **1932,** *11* (1), 93–170. DOI:10.1021/cr60038a002

19. Gao, Wei; Sammes, Nigel, M. An Introduction to Electronic and Ionic Materials. *World Sci.* **1999,** 261. ISBN 978-981-02-3473-7 (accessed Dec 3, 2017).

20. Souquet, J.. Electrochemical Properties of Ionically Conductive Glasses. *Solid State Ionics* **1981,** *5*, 77–82. DOI:10.1016/0167-2738(81)90198-3.

21. Cotton, F. Albert; Wilkinson, Geoffrey. *Advanced Inorganic Chemistry, 4th ed.; Wiley: New York, 1980.* ISBN 0-471-02775-8.

22. Ralph , H. P.; William, S. H.; Geoffrey, F. H.; Jeffry, D. M. *General Chemistry*. 9th ed.; Pearson Education, Inc.: New Jersey, 2007; pp 500, 513–515.

23. Combs, L. *Lattice Energy;* Chemistry Wikipedia, 1999.

24. Housecroft, C. E.; Alan, G. Sharpe. *Inorganic Chemistry*, 3rd ed.; Pearson Education Limited: England, 2008; pp 174–175.

25. Lang, P. F.; Smith, B. C. Ionic Radii for Group 1 and Group 2 Halide, Hydride, fluoride, Oxide, Sulfide, Selenide and Telluride Crystals. *Dalton Trans.* **2010,** *39* (33), 7786–7791. DOI:10.1039/C0DT00401D. PMID 20664858.

26. Pauling, L. *The Chemical Bond*; Cornell University: Ithaca NY, 1960; pp 150–15.

27. Cotton, F. A.; Wilkinson, G. *Advanced Inorganic Chemistry*, 3rd ed.; Wiley: London, 1972; p 50.

28. Shannon, R. D.; Prewitt, C. T. *Acta. Crystallog.* **1969,** *B25*, 925–926.

29. Adams, D. M. *Inorganic Solids*; Wiley: London, 1974; pp 35–37.

30. Järvinen, M.; Inkinen, O. *Physica Status Solidi (b)* **1967,** *21*, 127–135.

31. Berry, R. S.; Rice, S. A.; Ross, J. *Physical Chemistry*, 2nd ed.; Oxford University Press: Oxford, 2000; p 331.

32. Greenwood, N. N.; Earnshaw, A. *Chemistry of the Elements*; Pergamon: Oxford, 1984; p 92.
33. Johnson, D. A. *Some Thermodynamic Aspects of Inorganic Chemistry*; Cambridge University Press: London, 1968; p 23.
34. Greenwood, N. N.; Earnshaw, A. *Chemistry of the Elements*; Pergamon: Oxford, 1984; p 9.
35. Cheetham, A. K.; Day, P. *Solid State Chemistry*; Clarendon Press: Oxford, 1992.
36. Jenkins, H.; Donald, B. Thermodynamics of the Relationship Between Lattice Energy and Lattice Enthalpy. *J. Chem. Educ.* **2005,** *82*, 950–952 (Coventry, West Midlands, UK: University of Warwick).
37. Ladd, Mark. *Crystal Structures: Lattices and Solids in Stereoview*; Chichester: Horwood, 1999.
38. Ladd, Mark. *Chemical Bonding in Solids and Fluids*; Chichester: Horwood, 1994.
39. Suzuki, Takashi. *Free Energy and Self-interacting Particles*; Boston: Birkhauser, 2005.
40. *Brown, I. David. The Chemical Bond in Inorganic Chemistry: The Bond Valence Model (Reprint ed.);* Oxford University Press: *New York, 2002.* ISBN 0-19-850870-0.
41. Johnson. *The Open University: RSC. In Metals and Chemical Change (1. publ. ed.), David, Ed.;* Royal Society of Chemistry: *Cambridge, 2002.*
42. Cotton, F. Albert*;* Wilkinson, Geoffrey. *Advanced Inorganic Chemistry, 4th ed.; Wiley: New York, 1980,* ISBN 0-471-02775-8.
43. Johnson, D. A. *Metals and Chemical Change*; Royal Society of Chemistry: Cambridge, 2002.
44. Cotton, F. Albert *Advanced Inorganic Chemistry*; Wiley: New York, 1999.
45. David, A. J. *Metals and Chemical Change*; Open University, Royal Society of Chemistry, 2002. ISBN 0-85404-665-8.
46. Pauling, Linus. *The Nature of the Chemical Bond and the Structure of Molecules and Crystals: An Introduction to Modern Structural Chemistry*; Cornell University Press, 1960. ISBN 0-801-40333-2. DOI:10.1021/ja01355a027.
47. Pauling, L. *The Nature of the Chemical Bond*, 3rd ed.; Oxford University Press: USA, 1960; pp 98–100.
48. IUPAC, *Compendium of Chemical Terminology*, 2nd ed. (the "Gold Book"); 1997. Online corrected version: (2006) "Bond-dissociation energy" RSC Publishing: Cambridge, UK.
49. *Blanksby, S. J.; Ellison, G. B. Bond Dissociation Energies of Organic Molecules. Acc. Chem. Res.* **2003,** *36 (4): 255–263.* DOI:10.1021/ar020230d. PMID 12693923.
50. *Darwent, B. deB. Bond Dissociation Energies in Simple Molecules (PDF), NSRDS-NBS 31;* U. S. National Bureau of Standards*: Washington, DC, 1970,* LCCN 70602101.
51. *Morrison, R. T.; Boyd, R. N. Organic Chemistry, 4th ed.; Allyn & Bacon: Boston, 1983.* ISBN 0-205-05838-8.
52. Lehninger, A. L.*; Nelson, D. L.; Cox, M. M. Lehninger Principles of Biochemistry, 4th ed., Freeman.W. H., Ed.; 2005; p 48.* ISBN 978-0-7167-4339-2 (accessed May 20, 2016)*.*
53. *Schmidt-Rohr, K. Why Combustions Are Always Exothermic, Yielding About 418 kJ per Mole of O2. J. Chem. Educ.* **2015,** *92, 2094–2099.* Bibcode:2015JChEd..92.2094S. DOI:10.1021/acs.jchemed.5b00333.

CHAPTER 6

PHTHALONITRILE DERIVATIVES CONTAINING DIFFERENT HETEROCYCLIC GROUPS AS NEW CORROSION INHIBITORS FOR IRON(110) SURFACE

EBRU YABAŞ[1], LEI GUO[2], ZAKI S. SAFI[3], and SAVAŞ KAYA[4,*]

[1]*Department of Chemistry and Chemical Processing Technologies, Cumhuriyet University Imranlı Vocational School, 58980 Sivas, Turkey*

[2]*School of Material and Chemical Engineering, Tongren University, Tongren 554300, P.R. China*

[3]*Department of Chemistry, Faculty of Science, Al Azhar University-Gaza, Gaza City, P.O. Box 1277, Palestine*

[4]*Department of Chemistry, Faculty of Science, Cumhuriyet University, 58140 Sivas, Turkey*

Corresponding author. E-mail: savaskaya1989@gmail.com

ABSTRACT

It is important to note that iron is one of the most widely used metals in the industry. In the present study, we theoretically investigated corrosion inhibition and adsorption properties of some pyridine (*Inh-1*; synthesized in this study), imidazole[1] (*Inh-2*), and oxadiazole[2] (*Inh-3*) substituted phthalonitrile derivatives against the corrosion of iron metal with the help of quantum chemical calculations, molecular dynamics simulations approach, and principal component analysis. Calculated binding energies, quantum chemical parameters like HOMO and LUMO energies, chemical hardness, softness, electronegativity, HOMO–LUMO energy gap, proton affinity, dipole moment, electrophilicity, nucleophilicity, and principal component analysis performed showed that *Inh-2* is a better corrosion inhibitor compared to others and corrosion inhibition efficiency ranking for studied molecules can be given as *Inh-2 > Inh-1 > Inh-3*.

6.1 INTRODUCTION

In the industrial sector, one of the serious problems is the corrosion of metals and alloys. The corrosion of metals causes a huge financial loss.[3,4] The best method to protect the metals against corrosion is use of the corrosion inhibitor. Organic cyclic molecules containing oxygen, nitrogen, and/or sulfur atoms in their molecular structure are widely used as corrosion inhibitor and such molecules are very effective in terms of the prevention of the corrosion. The heteroatom and nature of the substituent play an important role in the adsorption process. The interactions between an organic inhibitor and a metal surface are principally physical adsorption and/or chemisorptions.[5–8]

In recent years, corrosion inhibition studies have focused on organic compound derivatives such as imidazole,[9–12] pyridine,[13–15] and oxadiazole.[16,17] These compound groups which have excellent ligand properties, the ability to form stable metal complexes, and environmental friendly are very promising as organic corrosion inhibitors. In the literature, there are a limited number of studies on corrosion inhibition of substituted phthalonitrile derivatives.[18–20] For example, the effect of 4-(2-diethylamino-ethylsulfanyl)-phthalonitrile and 4,5-bis (hexylsulfonyl)-phthalonitrile in the NaCl solution on the corrosion of stainless steel corrosion was investigated by Sezer et al.[18] The result of this study shows that there is electrostatic interaction between the metal surface and the phthalonitrile molecules. In another study, 4-(2-aminophenylthio)-5-(5-mercaptopentylthio)phthalonitrile was used to protect the copper surface from chloride corrosion by Arslan et al.[19] As a result, it has been determined that 4-(2-aminophenylthio)-5-(5-mercaptopentylthio)phthalonitrile films have predominantly anti-corrosion properties. In our work, corrosion inhibition properties of imidazole, pyridine, and oxadiazole substituted phthalonitrile derivatives were theoretically investigated and results showed that the synthesized molecules will be effective against the corrosion of iron metal.

6.2 MATERIAL AND METHODS

6.2.1 CHEMICALS

Inh-2[1] and *Inh-3*[2] were synthesized according to the related literatures. The synthesis of the new *Inh-1* compound is summarized below. The synthesized compounds were characterized by ^{1}H-NMR, ^{13}C-NMR, FT-IR, UV–Vis, and elemental analysis. All reactions were carried out in a nitrogen atmosphere. The solvents used in the reactions were prepared in the literature using drying techniques.[21] ^{1}H-NMR and ^{13}C-NMR spectra were recorded on a JEOL Resonance ECZ400S spectrometer. PerkinElmer Spectrum100 FT-IR Spectrometer was used for IR measurements (ATR). UV–Vis measurements were conducted with on a Shimadzu UV-1800 UV spectrophotometer. Melting point was determined with an Electrothermal 9100 digital melting point apparatus.

6.2.2 COMPUTATIONAL DETAILS

6.2.2.1 MOLECULAR DYNAMIC SIMULATIONS APPROACH

Adsorption process of the inhibitors compounds on the iron surface is investigated by MD simulation using Material Studio™ software 8.0 (from BIOVIA Inc.). In this present investigation, we have chosen Fe(110) surface for simulation. In this simulation process, interaction between inhibitors and iron surface are carried out in a simulation box of (22.3 × 22.3 × 45.1 Å) with periodic boundary condition in order to avoid any arbitrary boundary effects. Here, we have used five layers of iron atoms as it provides sufficient depth to overcome the issue related to cutoff radius in this case. In this investigation, simulation box is created by a two layer. The lower layer contains an iron slab and the upper layer is the solution slab which contains 600 H_2O and 1 inhibitor molecule. After construction of the simulation box, dynamics simulation is carried out using the condensed phase optimized molecular potentials for atomistic simulation studies (COMPASS) force field. COMPASS is a highly accepted and authentic ab initio force field that enables accurate prediction of nature for a good many number of chemical entities. The MD simulation is performed at 298.0 K using canonical ensemble (NVT) with a time step of 1.0 fs and a simulation time of 800 ps.[22–24]

The adsorption energy (E_{ads}) of the inhibitor molecule on the Fe(110) surface is calculated by the following formula:

$$E_{ads} = E_{total} - \left(E_{surf+H_2O} + E_{inh} \right), \tag{6.1}$$

where E_{total} is the total energy of the simulation system, E_{surf+H_2O} is the energy of the iron surface together with H_2O molecules, and E_{inh} is the energy of the free inhibitor molecule.

The binding energy of the inhibitor molecule is negative value of adsorption energy is as follows. Obviously, a larger value of $E_{binding}$ implies the corrosion inhibitor combines with the iron surface more easily and tightly, therefore higher and spontaneous inhibitive performance by the inhibitor molecules on the iron surfaces was observed.[24]

$$E_{binding} = -E_{ads}. \tag{6.2}$$

6.2.2.2 THE EQUATIONS USED TO CALCULATE THE QUANTUM CHEMICAL PARAMETERS

Quantum chemical parameters like hardness, softness, electronegativity, electrophilicity, nucleophilicity, dipole moment, chemical potential, proton affinity, the highest occupied molecular orbital (HOMO) energy, the lowest unoccupied molecular orbital (LUMO) energy, and HOMO–LUMO energy gap are widely used in the prediction of electron donating or electron accepting powers of chemical species. In the conceptual density functional theory (CDFT)[24,25] that provided important contributions to the development of quantum chemistry, aforementioned quantum chemical parameters are described as below as the derivative of electronic energy (E) with respect to number of electron (N) at a constant external potential $v(r)$:[26]

$$\chi = -\mu = -\left(\frac{\partial E}{\partial N}\right)_{\upsilon(r)}, \tag{6.3}$$

$$\eta = \frac{1}{2}\left(\frac{\partial \mu}{\partial N}\right)_{\upsilon(r)} = \frac{1}{2}\left(\frac{\partial^2 E}{\partial N^2}\right)_{\upsilon(r)}. \tag{6.4}$$

It is apparent from eq 6.3 that electronegativity is the negative of the chemical potential. Applying the finite differences method to the equations given above, Pearson and Parr[24,27] derived the following equations depending on ground state ionization energy (I) and electron affinity (A) values to calculate the hardness (η), electronegativity (χ), and chemical potential. Here it should be noted that softness (σ) is given as the multiplicative inverse of the hardness: ($\sigma = 1/\eta$),[24,28-30]

$$\text{where } \eta = \frac{I-A}{2} \tag{6.5}$$

and

$$\chi = -\mu = \frac{I+A}{2}. \tag{6.6}$$

Ionization energy and electron affinity values of molecules can be predicted in the light of Koopmans theorem.[24,31] According to mentioned theorem, the negative value of the highest occupied and lowest unoccupied molecular orbital energy corresponds to ionization energy and electron affinity, respectively ($-E_{HOMO} = I$ and $-E_{LUMO} = A$). In that case, within the framework of the theory, chemical hardness, chemical potential and electronegativity can be calculated with the help of the following equations:

$$\mu = -\chi = \frac{E_{LUMO} + E_{HOMO}}{2} \tag{6.7}$$

and

$$\eta = \frac{E_{LUMO} - E_{HOMO}}{2}. \qquad (6.8)$$

Proton affinity (PA) is one of the useful parameters used in corrosion inhibition studies. It should be noted that there is a remarkable relationship between gas phase basicity and proton affinity. Many researchers showed that effective corrosion inhibitors have more negative PA values compared to others. In this study, to predict the proton affinities of studied molecules, the following formula is used:

$$PA = E_{(pro)} - (E_{(non\text{-}pro)} + E_{H^+}), \qquad (6.9)$$

where, $E_{non\text{-}pro}$ and E_{pro} are the energies of the non-protonated and protonated inhibitors, respectively. E_H^+ is the energy of H^+ ion and was calculated as:

$$E_{H^+} = E_{(H_3O^+)} - E_{(H_2O)}. \qquad (6.10)$$

Parr and coworkers proposed the following formulas to calculate the electrophilicity index (ω) and nucleophilicity (ε) depending on electronegativity and chemical hardness. Nucleophilicity is defined as the multiplicative inverse of the electrophilicity:[24,27,32,33]

$$\omega = \chi^2 / 2\eta \qquad (6.11)$$

and

$$\varepsilon = 1 / \omega. \qquad (6.12)$$

6.2.3 SYNTHESIS OF 4[(2-METHOXYPYRIDINE)-5-OXO] PHTHALONITRILE (Inh-1)

4-hydroxyphthalonitrile (1.0 g, 5.78 mmol) and 2-methoxy-5-nitropyridine (1.6 g, 10.40 mmol) were dissolved in DMF (40 mL). A light-yellow solution was added K_2CO_3 (4.0 g, 28.90 mmol) for 2 h in the nitrogen atmosphere. Then the mixture was stirred at 80°C for 72 h. At the end of the reaction, the dark red mixture was poured into the water. The solid phase was filtered and dried. The crude product was dissolved in DMSO and the insoluble fraction removed. The solution was concentrated and precipitated with ether. Then it was filtered and dried in vacuum. The yellow solid was recrystallized in EtOH. The pale-yellow product is soluble in THF, DMF, and DMSO. Yield 580.0 mg (33%). Mp: 157°C. ^{1}H-NMR (400 MHz, DMSO-d$_6$, 25°C): δ = 7.9 (s, 1H, Ar-H); 7.1 (d, 2H, Ar-H); 6.3-6.2 (m, 3H, Ar-H); 2.9 (s, 3H, O-CH$_3$). ^{13}C-NMR (100 MHz, DMSO-d$_6$, 25°C): δ=168.02; 146.10; 141.38; 135.44; 132.71; 131.35; 129.11; 128.54; 113.86; 113.20; 112.02; 57.15. UV–Vis (DMSO) λ$_{max}$/nm (log ε, dm^3 mol^{-1} cm^{-1}) 361 (3.72), 307 (4.16). IR (ATR) υ (cm^{-1}) 3324; 3041; 2224; 1657; 1564; 1528; 1490; 1429; 1342; 1249; 1081; 866; 831; 735; 723. Anal. Calc. for C$_{14}$H$_9$N$_3$O$_2$: C 66.93; H 3.61; N 16.72%, found: C 67.04; H 3.67; N 16.86%.

6.3   RESULTS AND DISCUSSION

6.3.1   SYNTHESIS

Scheme 6.1 shows the synthesis route of compound *Inh-1* and the chemical structure of compounds *Inh-2*, *Inh-3*. 4[(2-methoxypyridine)-5-oxo]phthalonitrile (*Inh-1*) was synthesized by a substitution reaction of 4-hydroxyphthalonitrile and 2-methoxy-5-nitropyridine. The synthesized compound *Inh-1* was characterized with ^{1}H-NMR, ^{13}C-NMR, FT-IR, UV–Vis, spectra, and elemental analysis.

In the ^{1}H-NMR spectrum of compound *Inh-1*, the aliphatic methyl proton peaks were observed at 2.9 ppm while the aromatic proton peaks were observed between 7.9 and 6.2 ppm. The integral ratios of these peaks confirm that the desired compound is formed. On the other hand, the disappearance of the –OH group proton peak of 4-hydroxyphthalonitrile also supports the formation of the compound. ^{13}C-NMR spectrum of compound *Inh-1* supports the structure of the obtained compound. FT-IR and elemental analysis results are also compatible with the structure of the compound *Inh-1*.[34,35]

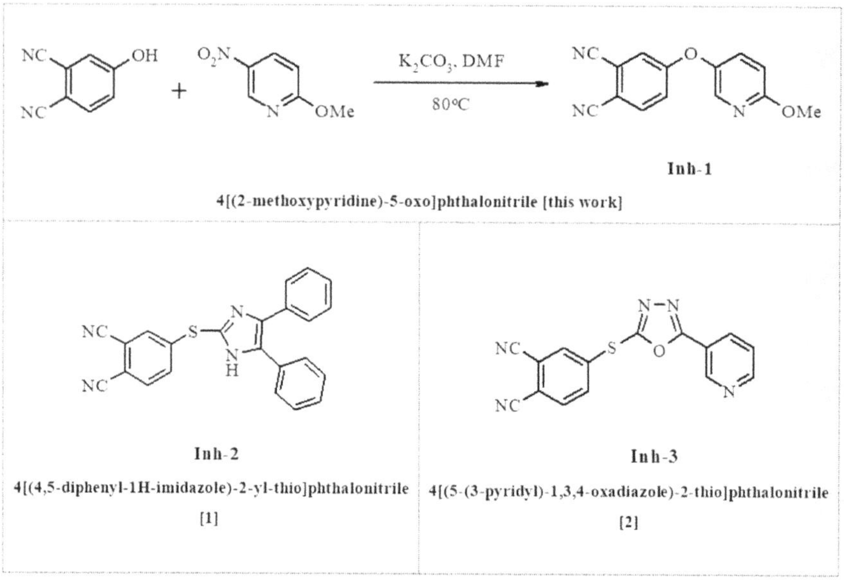

SCHEME 6.1 Synthesis of *Inh-1* and chemical structures of *Inh-1*, *Inh-2*, and *Inh-3*.

6.3.2 DFT ANALYSIS OF COMPOUNDS

It is well known that Frontier molecular orbital (FMO) theory is useful in predicting adsorption centers of the inhibitor molecules responsible for the interaction with surface metal atoms. Furthermore, the HOMO electron density surface reveals the electron rich sites of the molecule from which electrons could be donated to suitable vacant orbitals of accepting species (the metal). The LUMO electron density surface indicates the electron deficient regions of the molecule. These sites have higher chances of accepting electrons from appropriate occupied orbitals of donor species (the metal) during retro-bonding interactions between inhibitor and metal.[36-40] Figure 6.1 shows the optimized molecular structures, electron density surfaces of the HOMO, LUMO, and the electrostatic potential map (ESP) of the molecules under study (*Inh-1*, *Inh-2*, and *Inh-3*).

As shown in Figure 6.1, the HOMO electron density of *Inh-1* is principally distributed over the entire molecule, which indicates that these inhibitors can donate electron to the vacant orbitals of the metal. These results indicate that the HOMO of the *Inh-1* is mainly π-orbitals and the electron densities are distributed over the entire parts of the molecule.

Whereas, the HOMO of *Inh-2* is mainly localized over the imidazole and the phthalonitrile rings, which might be due to the presence of sulfur and nitrogen atoms with π-electrons of imidazol and phenyl rings. On the other hand, the HOMO of *Inh-3* is essentially localized on the pyrazolthiol part of the molecule, which might be due to the presence of S atom and pyrazol ring. The atoms of phthalonitrile and pyridine parts are not considerably involved in this distribution. This is due to their electron donating effects to the pyrazolthiol part. The LUMO of *Inh-1* is shown to be distributed overall the entire compound and the LUMO of *Inh-2* is mainly distributed over the mercapto phthalonitrile part of the molecule, while the LUMO of *Inh-3* is mainly located over the phthalonitrile part. Based on the above discussion, the electron density distributions of the FMOs suggest that the diphenyl imidazole part of *Inh-2* and the pyrazolthiol part of *Inh-3* are not likely to participate in the charge-sharing relationships between the inhibitor molecules and the metal.

Quantum chemical parameters are obtained from the calculations, which are responsible for the inhibition efficiency of inhibitors such as the energies of highest occupied molecular orbital (E_{HOMO}), energy of lowest unoccupied molecular orbital (E_{LUMO}), energy gap (ΔE), dipole moment (D), electronegativity (χ), electron affinity (A), global hardness (η), softness (σ), ionization potential (I), electrophilicity (ω), nucleophilicity (ε), the total energy (TE), and the fraction of electrons transferred from the inhibitor to copper surface (ΔN), are collected in Tables 6.1–6.4. Tables 6.1 and 6.2 represent gas and water phase results of the non-protonated species. Meanwhile, Tables 6.3 and 6.4 represent gas and water phase results of the protonated species.

As known, during the adsorption of the neutral organic molecules at the metal surface, an electron transfer takes place. In addition, the energy levels of the FMOs are significant for this transfer. Therefore, E_{HOMO} is often related to the molecule's electron-donating ability. High values of E_{HOMO} may indicate that a molecule is possibly raised to donate electrons to an unoccupied d-orbital of acceptor metals that are, of course, suitable to accept using low energy and empty molecular orbitals, and the higher the inhibition efficiency. As a result of this, the E_{LUMO} attributes to the suitability of the molecule to accept electrons from the filled 4f orbitals of the metal. For an electron to be accepted by a molecule, it can be assumed that the lower the value of the E_{LUMO}, the higher the possibility that the electron will be accepted. That is to say, a less negative value of E_{LUMO}

increases the effectiveness of the inhibition. This may indicate that the lower values of E_{LUMO} notify a better chance of electron accepting ability of an inhibitor during backward donation from the metal. Furthermore, the energy difference ($\Delta E = E_{LUMO} - E_{HOMO}$) between LUMO and HOMO is defined as the minimum energy required to excite an electron from the last occupied orbital in the molecule (HOMO). ΔE values are correlated to inhibition efficiencies determined experimentally for all the inhibitors. Generally, lower value of the energy gap, ΔE suggests higher reactivity and higher inhibition efficiency for a molecule.[24,41,31–33]

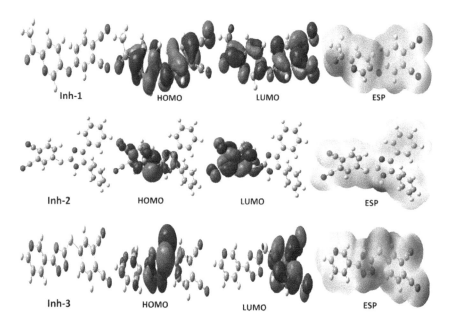

FIGURE 6.1   (See color insert.) The optimized structures, HOMOs, LUMOs, and electrostatic potential structures of non-protonated inhibitor molecules using DFT/6-311++G** calculation level.

Based on our calculated results (Tables 6.1–6.4), the results of E_{HOMO} and ΔE indicate that it is possible to get better performance with *Inh-2* as a corrosion inhibitor than the other inhibitors and the molecules under study follow the trend *Inh-2 > Inh-3 > Inh-1*. Also, *Inh-2* is the least negative E_{LUMO} value, reflecting the effective inhibition of *Inh-2* over the other inhibitors.

The computed values of the electronegativities of the inhibitors under probe (χ) are also mentioned in Tables 6.1–6.4. As it is well known, the χ values show the tendency of an atom to attract the shared pair of an electron toward itself. Consequently, a molecule with lower electronegativity (χ) may show higher inhibition efficiency. Based on or tabulated results in Tables 6.1–6.4, the electronegativity of the tested molecules follows the order *Inh-2 < Inh-3 < Inh-1*. On the other hand, the absolute hardness, η, determines both the stability and reactivity of a molecule, which suggests the resistivity of an inhibitor for the physical adsorption process. Soft molecules with small energy gaps are far more reactive than hard ones with large energy gaps, as they could readily offer electrons to an acceptor. An inspection of Tables 6.1–6.4 indicates that the absolute hardness of the investigated inhibitors are in the order—*Inh-2 < Inh-3 < Inh-1*, which follows the same trend of the decreasing of the energy gap. These results indicate that *Inh-2* is the most anticorrosive inhibitors among all the considered inhibitors.

The electronegativity values of *Inh-2* calculated at B3LYP/6-31++G level of theory are lower than that of iron, suggesting electron flow from the HOMO of the inhibitor toward the vacant 3d orbitals of Fe to be more significant than that from the filled 4s orbital of Fe to LUMO of the inhibitors (back-donation). In contrast, the electronegativity values of the other two inhibitors (*Inh-1* and *Inh-3*) are found to be higher than that of iron, which suggest a possible electron flow from 4s orbital of Fe to LUMO of the inhibitors.

The results presented in Tables 6.3 and 6.4 indicate that electrophilicity values (ω) are significantly higher for the protonated form of the inhibitors under study, suggesting that in the protonated form, the selected inhibitors have the highest capacity to accept electrons from the partially filled d-orbital of iron. These results are in agreement with those reported in the literature.

Comparison of Tables 6.1 and 6.2 clearly shows that the energy gaps (ΔE) values of the neutral solvated species calculated at B3LYP/6-31++G level of theory are lower than that of those in gas phase. For example, the energy gap (ΔE) value of *Inh-2* in gas phase is 0.0365 eV higher than in the aqueous medium. Furthermore, the hardness value of *Inh-2* is found to be 0.0182 eV higher than that in the aqueous medium. These results indicate that the inhibitors under study in aqueous medium are more reactive than in the gas phase. Additionally, the cationic species of the molecules under study were calculated with the same level of theory as neutral molecules. As seen from Tables 6.3 and 6.4, the softness (σ) values of cationic species of the investigated molecules are higher than the neutral counterparts. For example, the softness (σ) of *Inh-2* calculated at the B3LYP/6-31++G level are 0.550 for neutral *Inh-2*, whereas 1.478 for protonated *Inh-2*. Because of the softer molecule being more reactive, the protonated species interact with the metal surface more easily. Furthermore, the energy gap (ΔE) values of the protonated species are lower than the neutral ones, reflecting a higher reactivity of the inhibitors in the protonated species. These results agree with those reported in the literature.[24,41]

Pearson[42] proposed that the fraction of electrons transferred from the inhibitor molecule to the metallic surface is given by

$$\Delta N = \frac{\chi_{Fe} - \chi_{inh}}{\left[2\left(\eta_{Fe} + \eta_{inh}\right)\right]}, \tag{6.13}$$

where χ_{Fe} and χ_{inh} are the absolute electronegativity of metal and the inhibitor molecule, respectively, and η_{Fe} and η_{inh} correspond to the absolute hardness of metal and the inhibitor molecule, respectively. In order to calculate the transfer of electrons between iron and inhibitor molecule, for example, a theoretical value of 7 eV for the electronegativity of bulk iron χ_{Fe} and global hardness $\eta_{Fe} = 0$ are commonly used, by assuming that for a bulk metal $I = A$. It should be noted that the ΔN values do not indicate exactly the number of electrons leaving the donor and entering the acceptor molecule. The expression "electron-donating ability" has been suggested to be more adequate than "number of electron transferred". According to Lukovits,[43] if $\Delta N < 3.6$, the inhibition efficiency of organic inhibitor increases with increasing electron donating ability at the metal surface.[24]

More recently, using of the work function (φ) for the electronegativity of a metal surface in eq 6.13 has been proposed by Kokalj[44] which, indeed, leads to a reasonable, estimated trend of charge transfer for molecular adsorbate-metallic surface interactions. By using the work function for χ_{Fe} ($\varphi_{Fe} = 4.5$) and setting η_{Fe} to zero, eq 6.13 can be written as:

$$\Delta N^{\phi}_{HSAB} = \frac{\chi_{Fe} - \chi_{inh}}{\left[2\left(\eta_{Fe} + \eta_{inh}\right)\right]} = \frac{\phi - \chi_{inh}}{2\eta_{inh}}. \tag{6.14}$$

The fraction of electrons transferred, ΔN and ΔN_{HSAB}^{ϕ} calculated at all the suggested levels in both gas and aqueous phase of the protonated and non-protonated species are given in Table 6.5.[24]

According to Table 6.5, values of ΔN and ΔN_{HSAB}^{ϕ} show that the inhibition efficiency resulting from electron donation agrees with Lukovits's study.[43] If ΔN and or $\Delta N_{HSAB}^{\phi} < 3.6$, the inhibition efficiency increases by increasing electron-donating ability of these inhibitors to donate electrons to the metal surface. Table 6.5 shows and based on the results at B3LYP/6-31++G level of theory, we found that gas phase ΔN value of the neutral species decreases in the order—*Inh-2* $\approx$ *Inh-3* < *Inh-2*, while ΔN_{HSAB}^{ϕ} follows the order—*Inh-3* < *Inh-1* < *Inh-2*. On the other hand, the calculated ΔN values of the neutral species in aqueous medium decreases in the order—*Inh-1* < *Inh-3* < *Inh-2*, while ΔN_{HSAB}^{ϕ} decreases in the order—*Inh-3* < *Inb-1* < *Inh-2*. Cationic species of the studied inhibitors in the gas phase show that ΔN decreases in the order—*Inh-1* < *Inh-3* < *Inh-2*, while ΔN_{HSAB}^{ϕ} decreases in the order—*Inh-2* < *Inb-3* < *Inh-1*. On the other hand, the two values decrease in the order—*Inh-1* < *Inh-3* < *Inh-2* in the aqueous medium. Thus, the highest fraction of electrons transferred is associated with the best inhibitor (*Inh-2*), while the least fraction is associated with the inhibitor that has the least inhibition efficiency (*Inh-1*).

The dipole moment values of the tested inhibitors are also listed in Tables 6.1–6.4. Previous studies suggested that the low value of dipole moment has often been associated with good inhibition properties. The direction of the dipole can be understood by considering the electrostatic potential (right panels of Fig. 6.1). Our dipole moments values are substantially increased for the investigated inhibitors ongoing from the gaseous to the aqueous phase, indicating an increase in the stability of the inhibitors due to the interaction with water. For example, at the B3LYP/6-31++G level of theory the dipole moments of the tested *Inh-1*, *Inh-2*, and *Inh-3* inhibitors are increased from 5.08, 9.4, and 6.6 Debye to 6.26, 12.12, and 8.56 Debye, respectively. The dipole moments order follows the trend—*Inh-2* > *Inh-3* > *Inh-1*, suggesting that *Inh-2* may be the best corrosion inhibitors among all the tested compounds. These results agree with global descriptors findings.

TABLE 6.1 Calculated Quantum Chemical Parameters for Non-protonated Molecules in Gas Phase.

Gas	E_{HOMO} (eV)	E_{LUMO} (eV)	I	A	ΔE	η	σ	X	PA	ω	ε	Dipole moment	Energy (eV)
HF/SDD LEVEL													
Inh-1	-10.13197	0.89771	10.13197	-0.89771	11.02968	5.51484	0.18133	4.61713	-1.33860	1.93277	0.51739	8.4531	-24105, 29771
Inh-2	-8.19940	0.96737	8.19940	-0.96737	9.16677	4.58339	0.21818	3.61601	-1.09916	1.42641	0.70106	9.8874	-40647, 69404
Inh-3	-9.79781	0.58042	9.79781	-0.58042	10.37823	5.18912	0.19271	4.60869	-0.70525	2.04660	0.48862	6.8378	-35813, 97307
HF/6-31G LEVEL													
Inh-1	-9.98720	1.18670	9.98720	-1.18670	11.17390	5.58695	0.17899	4.40025	-1.43185	1.73281	0.57710	8.2057	-24102, 68738
Inh-2	-8.02579	1.25908	8.02579	-1.25908	9.28487	4.64244	0.21540	3.38335	-1.12629	1.23288	0.81111	9.7863	-40645, 32935
Inh-3	-9.68842	0.82533	9.68842	-0.82533	10.51375	5.25687	0.19023	4.43155	-0.71931	1.86790	0.53536	6.8217	-35811, 51844
HF/6-31++G LEVEL													
Inh-1	-10.16897	0.54450	10.16897	-0.54450	10.71348	5.35674	0.18668	4.81224	-1.32828	2.16154	0.46263	8.0654	-24103, 29406
Inh-2	-8.20375	0.66723	8.20375	-0.66723	8.87098	4.43549	0.22545	3.76826	-0.97080	1.60070	0.62473	9.6348	-40646, 16638
Inh-3	-9.80271	0.58042	9.80271	-0.58042	10.38313	5.19157	0.19262	4.61114	-0.68171	2.04781	0.48833	7.0112	-35812, 28459
B3LYP/SDD LEVEL													
Inh-1	-7.29053	-2.92661	7.29053	2.92661	4.36392	2.18196	0.45830	5.10857	-1.44562	5.98028	0.16722	5.1521	-24254, 88556
Inh-2	-6.15853	-2.59109	6.15853	2.59109	3.56744	1.78372	0.56063	4.37481	-1.43977	5.36490	0.18640	9.5621	-40860, 02757
Inh-3	-7.32781	-2.85939	7.32781	2.85939	4.46842	2.23421	0.44759	5.09360	-1.11864	5.80626	0.17223	7.9333	-35988, 32152
B3LYP/6-31G LEVEL													
Inh-1	-7.12318	-2.64796	7.12318	2.64796	4.47522	2.23761	0.44691	4.88557	-1.51852	5.33355	0.18749	4.9296	-24252, 11477
Inh-2	-5.98737	-2.34564	5.98737	2.34564	3.64173	1.82086	0.54919	4.16650	-1.48981	4.76690	0.20978	9.2939	-40857, 42160
Inh-3	-7.18250	-2.59381	7.18250	2.59381	4.58869	2.29435	0.43585	4.88816	-1.16727	5.20716	0.19204	7.8437	-35985, 41227
B3LYP/6-31++G LEVEL													
Inh-1	-7.45299	-2.96470	7.45299	2.96470	4.48828	2.24414	0.44560	5.20884	-1.31974	6.04509	0.16542	5.0784	-24253, 04478
Inh-2	-6.29785	-2.66075	6.29785	2.66075	3.63710	1.81855	0.54989	4.47930	-1.33427	5.51652	0.18127	9.3537	-40858, 51558
Inh-3	-7.44673	-3.05613	7.44673	3.05613	4.39059	2.19530	0.45552	5.25143	-0.93478	6.28105	0.15921	6.5508	-35986, 53989

TABLE 6.2 Calculated Quantum Chemical Parameters for Non-protonated Molecules in Aqueous Phase.

Water	E_{HOMO} (eV)	E_{LUMO} (eV)	I	A	ΔE	η	σ	χ	PA	ω	ε	Dipole moment	Energy (eV)
HF/SDD LEVEL													
Inh-1	-9.81903	1.16520	9.81903	-1.16520	10.98423	5.49212	0.18208	4.32692	-1.33860	1.70446	0.58670	8.3529	-24106, 20894
Inh-2	-8.19532	0.99676	8.19532	-0.99676	9.19208	4.59604	0.21758	3.59928	-1.09916	1.40934	0.70955	11.7285	-40648, 44277
Inh-3	-9.56678	0.81472	9.56678	-0.81472	10.38150	5.19075	0.19265	4.37603	-0.70525	1.84460	0.54212	8.7167	-35814, 75249
HF/6-31G LEVEL													
Inh-1	-9.68461	1.41636	9.68461	-1.41636	11.10097	5.55049	0.18016	4.13412	-1.43185	1.53959	0.64952	8.4063	-24103, 55225
Inh-2	-	-	-	-	-	-	-	-	-	-	-	-	-
Inh-3	-9.48515	1.03186	9.48515	-1.03186	10.51701	5.25851	0.19017	4.22664	-0.71931	1.69863	0.58871	8.6625	-35812, 25533
HF/6-31++G LEVEL													
Inh-1	-9.80815	1.08792	9.80815	-1.08792	10.89607	5.44803	0.18355	4.36011	-1.32828	1.74472	0.57316	8.7776	-24104, 21425
Inh-2	-8.17736	0.95622	8.17736	-0.95622	9.13357	4.56679	0.21897	3.61057	-0.97080	1.42729	0.70063	11.7169	-40646, 91114
Inh-3	-9.57413	0.80057	9.57413	-0.80057	10.37469	5.18735	0.19278	4.38678	-0.68171	1.85488	0.53912	9.0217	-35813, 05816
B3LYP/SDD LEVEL													
Inh-1	-7.17271	-2.65286	7.17271	2.65286	4.51985	2.25992	0.44249	4.91278	-1.44562	5.33988	0.18727	6.4656	-24255, 67712
Inh-2	-6.11200	-2.53558	6.11200	2.53558	3.57642	1.78821	0.55922	4.32379	-1.43977	5.22733	0.19130	12.1365	-40860, 67834
Inh-3	-7.12427	-2.82647	7.12427	2.82647	4.29780	2.14890	0.46535	4.97537	-1.11864	5.75976	0.17362	8.2284	-35988, 95859
B3LYP/6-31G LEVEL													
Inh-1	-7.00318	-2.39272	7.00318	2.39272	4.61046	2.30523	0.43380	4.69795	-1.51852	4.78709	0.20890	6.0798	-24252, 82934
Inh-2	-5.91580	-2.32523	5.91580	2.32523	3.59057	1.79529	0.55701	4.12052	-1.48981	4.72868	0.21148	11.2192	-40858, 00523
Inh-3	-6.99583	-2.60497	6.99583	2.60497	4.39086	2.19543	0.45549	4.80040	-1.16727	5.24813	0.19054	8.0716	-35986, 10097
B3LYP/6-31++G LEVEL													
Inh-1	-7.25407	-2.69504	7.25407	2.69504	4.55903	2.27952	0.43869	4.97455	-1.31974	5.42795	0.18423	6.2556	-24253, 86997
Inh-2	-6.17540	-2.57476	6.17540	2.57476	3.60064	1.80032	0.55546	4.37508	-1.33427	5.31609	0.18811	12.1232	-40859, 23952
Inh-3	-7.19366	-2.84334	7.19366	2.84334	4.35032	2.17516	0.45974	5.01850	-0.93478	5.78931	0.17273	8.5599	-35987,20312

TABLE 6.3   Calculated Quantum Chemical Parameters for Protonated Molecules in Gas Phase.

	E_{HOMO} (eV)	E_{LUMO} (eV)	I	A	ΔE	η	σ	χ	ω	ε	Energy (eV)
HF/SDD LEVEL											
Inh-1	−12.98728	−4.27494	12.98728	4.27494	8.71234	4.35617	0.22956	8.63111	8.55064	0.11695	−24113, 99630
Inh-2	−10.09360	−3.56200	10.09360	3.56200	6.53160	3.26580	0.30620	6.82780	7.13743	0.14011	−40656, 15320
Inh-3	−11.69201	−3.89752	11.69201	3.89752	7.79449	3.89725	0.25659	7.79476	7.79503	0.12829	−35822, 03831
HF/6-31G LEVEL											
Inh-1	−12.83490	−4.14269	12.83490	4.14269	8.69220	4.34610	0.23009	8.48879	8.29015	0.12063	−24111, 47923
Inh-2	−9.94366	−3.43791	9.94366	3.43791	6.50575	3.25287	0.30742	6.69079	6.88109	0.14533	−40653, 81564
Inh-3	−11.60275	−3.75928	11.60275	3.75928	7.84347	3.92174	0.25499	7.68102	7.52193	0.13294	−35819, 59776
HF/6-31++G LEVEL											
Inh-1	−12.95191	−4.25290	12.95191	4.25290	8.69900	4.34950	0.22991	8.60240	8.50687	0.11755	−24111, 98235
Inh-2	−10.09142	−3.58594	10.09142	3.58594	6.50548	3.25274	0.30743	6.83868	7.18896	0.13910	−40654, 49719
Inh-3	−11.68221	−3.88745	11.68221	3.88745	7.79476	3.89738	0.25658	7.78483	7.77491	0.12862	−35820, 32629
B3LYP/SDD LEVEL											
Inh-1	−10.55211	−7.84810	10.55211	7.84810	2.70402	1.35201	0.73964	9.20011	31.30230	0.03195	−24263, 69118
Inh-2	−8.16702	−6.80698	8.16702	6.80698	1.36004	0.68002	1.47055	7.48700	41.21595	0.02426	−40868, 82734
Inh-3	−9.23371	−7.46714	9.23371	7.46714	1.76658	0.88329	1.13213	8.35042	39.47157	0.02533	−35996, 80016
B3LYP/6-31G LEVEL											
Inh-1	−10.40381	−7.69653	10.40381	7.69653	2.70728	1.35364	0.73875	9.05017	30.25380	0.03305	−24260, 99329
Inh-2	−8.01055	−6.69160	8.01055	6.69160	1.31895	0.65947	1.51636	7.35108	40.97085	0.02441	−40866, 27141
Inh-3	−8.99071	−7.35475	8.99071	7.35475	1.63596	0.81798	1.22252	8.17273	40.82832	0.02449	−35993, 93954
B3LYP/6-31++G LEVEL											
Inh-1	−10.62368	−7.88320	10.62368	7.88320	2.74048	1.37024	0.72980	9.25344	31.24495	0.03201	−24261, 72452
Inh-2	−8.25001	−6.89678	8.25001	6.89678	1.35323	0.67662	1.47794	7.57340	42.38467	0.02359	−40867, 20985
Inh-3	−9.36569	−7.48346	9.36569	7.48346	1.88223	0.94111	1.06257	8.42458	37.70719	0.02652	−35994, 83467

TABLE 6.4 Calculated Quantum Chemical Parameters for Protonated Molecules in Aqueous Phase.

	E_{HOMO} (eV)	E_{LUMO} (eV)	I	A	ΔE	η	σ	χ	ω	ε	Energy (eV)
HF/SDD LEVEL											
Inh-1	-10.45524	-0.32082	10.45524	0.32082	10.13442	5.06721	0.19735	5.38803	2.86459	0.34909	-24116, 90122
Inh-2	-8.29355	0.08109	8.29355	-0.08109	8.37464	4.18732	0.23882	4.10623	2.01336	0.49668	-40658, 74073
Inh-3	-9.62665	-0.09034	9.62665	0.09034	9.53631	4.76815	0.20972	4.85850	2.47527	0.40400	-35825, 03415
HF/6-31G LEVEL											
Inh-1	-10.32816	0.19973	10.32816	-0.19973	10.52790	5.26395	0.18997	5.06421	2.43603	0.41050	-24114, 28709
Inh-2	-8.11232	0.11347	8.11232	-0.11347	8.22579	4.11290	0.24314	3.99942	1.94454	0.51426	-40656, 33427
Inh-3	-9.54665	0.02531	9.54665	-0.02531	9.57195	4.78598	0.20894	4.76067	2.36775	0.42234	-35822, 51933
HF/6-31++G LEVEL											
Inh-1	-10.42368	-0.31838	10.42368	0.31838	10.10530	5.05265	0.19792	5.37103	2.85473	0.35030	-24114, 85989
Inh-2	-8.26743	0.02422	8.26743	-0.02422	8.29165	4.14582	0.24121	4.12160	2.04876	0.48810	-40657, 09086
Inh-3	-9.63155	-0.09687	9.63155	0.09687	9.53467	4.76734	0.20976	4.86421	2.48153	0.40298	-35823, 22485
B3LYP/SDD LEVEL											
Inh-1	-7.75422	-4.15875	7.75422	4.15875	3.59547	1.79773	0.55626	5.95648	9.86789	0.10134	-24266, 34567
Inh-2	-6.18356	-3.60935	6.18356	3.60935	2.57422	1.28711	0.77694	4.89646	9.31362	0.10737	-40871, 13145
Inh-3	-7.22250	-3.88418	7.22250	3.88418	3.33832	1.66916	0.59910	5.55334	9.23807	0.10825	-35999, 22558
B3LYP/6-31G LEVEL											
Inh-1	-7.63694	-3.99820	7.63694	3.99820	3.63874	1.81937	0.54964	5.81757	9.30106	0.10751	-24263, 59651
Inh-2	-5.98546	-3.47683	5.98546	3.47683	2.50864	1.25432	0.79725	4.73114	8.92267	0.11207	-40868, 49179
Inh-3	-7.10413	-3.74568	7.10413	3.74568	3.35846	1.67923	0.59551	5.42490	8.76283	0.11412	-35996, 46524
B3LYP/6-31++G LEVEL											
Inh-1	-7.96728	-4.16365	7.96728	4.16365	3.80364	1.90182	0.52581	6.06547	9.67229	0.10339	-24264, 42255
Inh-2	-6.23989	-3.61996	6.23989	3.61996	2.61993	1.30997	0.76338	4.92993	9.27664	0.10780	-40869, 52368
Inh-3	-7.28890	-3.87929	7.28890	3.87929	3.40961	1.70481	0.58658	5.58409	9.14534	0.10935	-35997, 39343

TABLE 6.5 The Fraction of Electrons Transferred, ΔN and ΔN^{ϕ}_{HSAB} of the Inhibitors Under Study.

| | ΔN | | | | ΔN^{ϕ}_{HSAB} | | | |
| | Non-protonated | | Protonated | | Non-protonated | | Protonated | |
	Gas	Water	Gas	Water	Gas	Water	Gas	Water
HF/SDD LEVEL								
Inh-1	0.216	0.243	−0.187	0.159	−0.011	0.016	−0.474	−0.088
Inh-2	0.369	0.370	0.026	0.346	0.096	0.098	−0.356	0.047
Inh-3	0.230	0.253	−0.102	0.225	−0.010	0.012	−0.423	−0.038
HF/6-31G LEVEL								
Inh-1	0.233	0.258	−0.171	0.184	0.009	0.033	−0.459	−0.054
Inh-2	0.390	–	0.048	0.365	0.120	–	−0.337	0.061
Inh-3	0.244	0.264	−0.087	0.234	0.007	0.026	−0.406	−0.027
HF/6-31++G LEVEL								
Inh-1	0.204	0.242	0.312	3.058	−0.029	0.013	0.245	0.213
Inh-2	0.364	0.371	0.522	−58.922	0.082	0.097	0.322	0.257
Inh-3	0.230	0.252	0.399	11.524	−0.011	0.011	0.272	0.225
B3LYP/SDD LEVEL								
Inh-1	0.433	0.462	−0.814	0.290	−0.139	−0.091	−1.738	−0.405
Inh-2	0.736	0.748	−0.358	0.817	0.035	0.049	−2.196	−0.154
Inh-3	0.427	0.471	−0.764	0.433	−0.133	−0.111	−2.180	−0.316
B3LYP/6-31G LEVEL								
Inh-1	0.472	0.499	−0.757	0.325	−0.086	−0.043	−1.681	−0.362
Inh-2	0.778	0.802	−0.266	0.904	0.092	0.106	−2.162	−0.092
Inh-3	0.460	0.501	−0.717	0.469	−0.085	−0.068	−2.245	−0.275
B3LYP/6-31++G LEVEL								
Inh-1	0.399	0.444	−0.822	0.246	−0.158	−0.104	−1.735	−0.412
Inh-2	0.693	0.729	−0.424	0.790	0.006	0.035	−2.271	−0.164
Inh-3	0.398	0.455	−0.757	0.415	−0.171	−0.119	−2.085	−0.318

6.3.3 MONTE CARLO ANALYSIS OF COMPOUNDS

One of the most useful tools used to explanation between inhibitor molecules and metal surface is molecular dynamics simulation approach.[45–47]

The most stable low energy adsorption configurations for *Inh-1*, *Inh-2*, and *Inh-3* molecules on Fe(110) surface are given in Figure 6.2. Binding energy is the negative value of adsorption energy. In Table 6.6, calculated binding energy and adsorption energy values for studied inhibitor molecules are given. It is apparent from the results given in the mentioned table that corrosion inhibition efficiency ranking of studied molecules can be given as *Inh-2* > *Inh-3* > *Inh-1*. The results obtained via molecular dynamics simulation approach are in good agreement with DFT data.

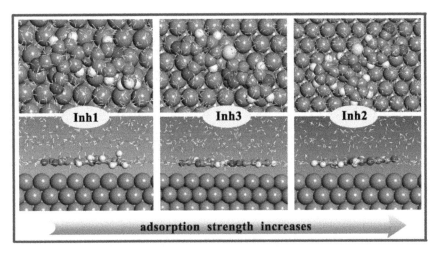

FIGURE 6.2 **(See color insert.)** Equilibrium adsorption configurations of inhibitors *Inh-1*, *Inh-2*, and *Inh-3* on Fe(110) surface obtained by MD simulations. Upper: top view and lower: side view.

TABLE 6.6 Output Obtained from MD Simulation for Adsorption of Inhibitors on Fe(110) Surface.

Systems	E_{ads} (kcal/mol)	$E_{binding}$ (kcal/mol)
Fe(110) + *Inh-1*	−187.7	187.7
Fe(110) + *Inh-2*	−247.8	247.8
Fe(110) + *Inh-3*	−199.5	199.5

6.3.4   PRINCIPAL COMPONENT ANALYSIS AND AGGLOMERATIVE HIERARCHICAL CLUSTER ANALYSIS

As it is well known that principal component analysis (PCA) involves a mathematical procedure that converts a number of possibly correlated variables into a smaller number of uncorrelated variables called principal components. The first principal component (PC1) accounts for as much of the variability in the data as possible, and each succeeding component (PC2, PC3, ...) accounts for as much of the remaining variability as possible. Data sets with many variables can be simplified through variable reduction and thereby, be more easily interpreted. In this study, the PCA was used to exploit the results, build statistical models and determine the appropriate quantum descriptors, which can be used to select the best inhibitor for the corrosion inhibition process. Actually, in this work, all calculated variables have been auto scaled to compare them on the same scale. Thereafter, PCA was used to reduce the number of variables and select the most relevant ones, that is, those responsible for the reactivity of the studied inhibitors. After performing many tests, a good separation is obtained between more active and less active inhibitor compounds using 11 variables: E_{HOMO}, I, ΔE_{gap}, χ, μ, η, σ, μ, ω, ΔN_{max}, $\Delta\psi$, and ΔE_{b-d}. As indicated from PCA results that the first two principal components (PC1 and PC2) describe all of the overall variance as follows: PC1 = 95.32 % and PC2 = 4.68 % in gas phase, and in aqueous medium, the PC1 = 87.70% and PC2 = 12.30%. The score plot of the variances is a reliable representation of the spatial distribution of the points for the data set studied after explaining almost all of the variances by the first two PCs.[24] In Figure 6.3, the most informative biplot plot for quantum chemical descriptors obtained in aqueous medium at B3LYP/6-311G level of theory for the neutral inhibitors is presented (PC1 vs. PC2). The same biplot was also examined for the results obtained in the gas phase (not shown here). The PCA results in the two phases are almost the same. It is evident from the figure that PC1 alone is responsible for the separation between more active inhibitor *Inh-2* and the less active inhibitors *Inh-1* and *Inh-3*, where PC1 > 0 for the more active compounds and PC1 < 0 for the less active one. Moreover, based on the biplot figure, the quantum chemical descriptors have been divided into parts (more active and less active). Based on the figure, the most important variables with PC1 > 1, which control the reactivity of the inhibitor reactivity, are E_{HOMO}, E_{LUMO}, ΔN_{max}, dipole moment (DP), ΔE_{b-d} and Q_{max} descriptors. In contrast, the less important variables with PC1 < 1 are ΔE_{gap}, ω, χ, η, and $\Delta\psi$. This finding has been also confirmed by the Agglomerative Hierarchical Cluster Analysis (AHCA; Fig. 6.2). In this figure, the horizontal lines represent the inhibitors and the vertical lines the similarity values between pairs of inhibitors, an inhibitor and a group of inhibitors and among groups of inhibitors. It is noticed, in agreement with the PCA results, that AHCA the studied inhibitors were grouped into two categories: More active inhibitor is *Inh-2* and less active ones are *Inh-1* and *Inh-3* (Figs 6.3 and 6.4).

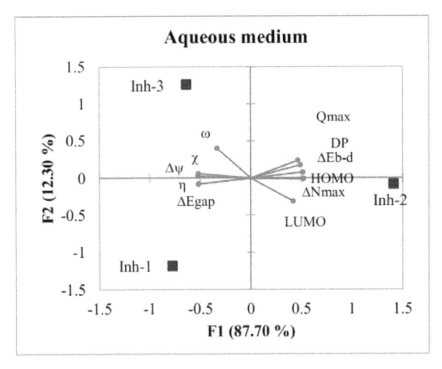

FIGURE 6.3 **(See color insert.)** PCA based on a correlation matrix combining quantum chemical descriptors (red circles) and the studied inhibitors (*Inh-1*, *Inh-2*, and *Inh-3*; blue squares).

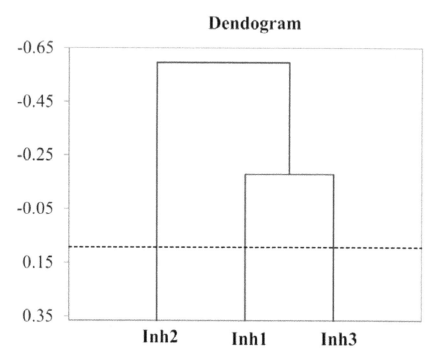

FIGURE 6.4 **(See color insert.)** Dendrogram obtained for the studied inhibitors (*Inh-1*, *Inh-2*, and *Inh-3*) in aqueous phase.

6.4 CONCLUSION

With the help of HF, DFT calculations, molecular simulation dynamics approach, and PCA, we theoretically analyzed the corrosion inhibition performances of synthesized phthalonitrile derivatives containing different heterocyclic groups against the corrosion of iron. In the light of quantum chemical calculations made for gas phase and aqueous solution, the results of molecular dynamics simulations approach and the results of PCA, the following conclusions can be presented:

- All theoretical data obtained in this study showed that the corrosion inhibition efficiency ranking for studied phthalonitrile derivatives containing different heterocyclic groups against the corrosion of iron can be given as: *Inh-2 > Inh-3 > Inh-1*.

- Synthesized phthalonitrile derivatives will be effective in terms of the prevention of the corrosion of iron metal.
- The binding energies calculated considering interactions between metal surface and inhibitor molecules are in good agreement with calculated quantum chemical parameters. According to calculated binding energies given in related table, the most effective inhibitor among studied molecules against the corrosion of iron is imidazole substituted phthalonitrile *Inh-2*.
- The results obtained and discussions made in this theoretical study are very important in terms of the design of new phthalonitrile derivatives in future studies.

KEYWORDS

- **density functional theory (DFT)**
- **phthalonitrile**
- **oxadiazole**
- **pyridine**
- **imidazole**
- **corrosion inhibitors**
- **Fe(110) surface**

REFERENCES

1. Yabaş, E.; Sülü, M.; Saydam, S.; Dumludağ, F.; Salih, B. Bekaroğlu, Ö. Synthesis, Characterization and İnvestigation of Electrical and Electrochemical Properties of İmidazole Substituted Phthalocyanines. *Inorg. Chim. Acta.* **2011,** *365,* 340–348.

2. Bağda, E.; Yabaş, E.; Bağda, E. Analytical Approaches for Clarification of DNA-Double Decker Phthalocyanine Binding Mechanism: As an Alternative Anticancer Chemotherapeutic. *Spectrochim. Acta Part A: Mol. Biomol. Spectrosc.* **2017,** *172,* 199–204.

3. Obot, I. B.; Macdonald, D. D.; Gasem, Z. M. Density Functional Theory (DFT) as a Powerful Tool for Designing New Organic Corrosion İnhibitors. Part 1: An Overview. *Corr. Sci.* **2015,** *99,* 1–30.

4. Li, X.; Zhang, D.; Liu, Z.; Li, Z.; Du, C.; Dong, C. Materials Science: Share Corrosion Data. *Nature* **2015,** *527,* 441–442.

5. Obot, I. B.; Kaya, S.; Kaya, C.; Tüzün, B. Density Functional Theory (DFT) Modeling and Monte Carlo Simulation Assessment of İnhibition Performance of Some Carbohydrazide Schiff Bases for Steel Corrosion. *Phys. E: Low-dimensional Syst. Nanostruct.* **2016**, *80*, 82–90.

6. Gupta, S. R.; Mourya, P.; Singh, M. M.; Shing, V. P. Structural, Theoretical and Corrosion İnhibition Studies on Some Transition Metal Complexes Derived from Heterocyclic System. *J. Mol. Struct.* **2017**, *1137*, 240–252.

7. Kaya, S.; Tüzün, B.; Kaya, C.; Obot, I. B. Determination of Corrosion İnhibition Effects of Amino Acids: Quantum Chemical and Molecular Dynamic Simulation Study. *J. Taiwan Inst. Chem. Eng.* **2016**, *58*, 528–535.

8. Wazzan, N. A.; Obot, I. B.; Kaya, S. Theoretical Modeling and Molecular Level İnsights into the Corrosion İnhibition Activity of 2-Amino-1, 3, 4-Thiadiazole and its 5-Alkyl Derivatives. *J. Mol. Liq.* **2016**, *221*, 579–602.

9. Mihajlović, M. B. P.; Radovanović, M. B. Tasić, Ž. Z. Antonijević, M. M. Imidazole Based Compounds as Copper Corrosion İnhibitors in Seawater. *J. Mol. Liquids.* **2017**, *225*, 127–136.

10. Kovačević, N.; Milošev, I.; Kokalj, A. How Relevant is the Adsorption Bonding of İmidazoles and Triazoles for their Corrosion İnhibition of Copper? *Corros. Sci.* **2017**, *124*, 25–34.

11. Singh, A.; Ansari, K. R.; Kumar, A.; Liu, W.; Songsong, C.; Lin, Y. Electrochemical, Surface and Quantum Chemical Studies of Novel İmidazole Derivatives as Corrosion İnhibitors for J55 Steel in Sweet Corrosive Environment. *J. Alloys Compd.* **2017**, *712*, 121–133.

12. Xhanari, K.; Finsgar, M. The First Electrochemical and Surface Analysis of 2-Aminobenzimidazole as a Corrosion İnhibitor for Copper in Chloride Solution. *New J. Chem.* **2017**, *41*, 7151–7161.

13. Dohare, P.; Quraishi, M. A.; Obot, I. B. A Combined Electrochemical and Theoretical Study of Pyridine-Based Schiff Bases as Novel Corrosion İnhibitors for Mild Steel in Hydrochloric Acid Medium. *J. Chem. Sci.* **2018**, *8*, 130–148.

14. Cruz-Borbolla, J.; Garcia-Ochoa, E.; Narayanan, J.; Maldonado-Rivas, P.; Pandiyan, T.; Vásquez-Pérez, J. M. Electrochemical and Theoretical Studies of the İnteractions of a Pyridyl-Based Corrosion İnhibitor with İron Clusters (Fe15, Fe30, Fe45, and Fe60). *J. Mol. Model.* **2017**, *342*, 23–38.

15. Han, P.; Li, W.; Tian, H.; Gao, X.; Ding, R.; Xiong, C.; Song, L.; Zhang, X.; Wang, W.; Chen, C. Comparison of İnhibition Performance of Pyridine Derivatives Containing Hydroxyl and Sulfhydryl Groups: Experimental and Theoretical Calculations. *Mater. Chem. Phys.* **2018**, *214*, 345–354.

16. Ammal, P. R.; Prajila, M.; Joseph, A. Effect of Substitution and Temperature on the Corrosion İnhibition Properties of Benzimidazole Bearing 1, 3, 4-Oxadiazoles for Mild Steel in Sulphuric Acid: Physicochemical and Theoretical Studies. *J. Environ. Chem. Eng.* **2018**, *6*, 1072–1085.

17. Ammal, P. R.; Prajila, M.; Joseph, A. Effective İnhibition of Mild Steel Corrosion in Hydrochloric Acid Using EBIMOT, a 1, 3, 4-Oxadiazole Derivative Bearing a 2-Ethylbenzimidazole Moiety: Electro Analytical, Computational and Kinetic Studies. *Egyptian J. Petroleum.* **2017**, 1–11.

18. Sezer, E.; Ustamehmetoğlu, B.; Altuntaş Bayır, Z.; Çoban, K.; Kalkan, A. Corrosion Inhibition Effect of 4-(2-Diethylamino-Ethylsulfonyl)-Phthalonitrile and 4,5-Bis(Hexylsulfonyl)-Phthalonitrile. *Int. J. Electrochem.* **2011.** DOI:10.4061/2011/235360.

19. Arslan, E.; Gürten, A. A.; Gök, H. Z.; Farsak, M. Electrochemical Study of Self-Assembled Aminothiol Substituted Phthalonitrile Layers for Corrosion Protection of Copper. *Chem. Sel.* **2017,** *2,* 8256–8261.

20. Derradji, M.; Ramdani, N.; Zhang, T.; Wang, J.; Gong, L.; Xu, X.; Lin, Z.; Henniche, A.; Rahoma, H. K. S.; Liu, W. Effect of Silane Surface Modified Titania Nanoparticles on the Thermal, Mechanical, and Corrosion Protective Properties of a Bisphenol-Abased Phthalonitrile Resin. *Prog. Organ. Coat.* **2016,** *90,* 34–43.

21. Armarego, W. L. F.; Chai, C. L. L. Purification of Laboratory Chemicals, 3rd ed; Butterworth/Heinemann: Tokyo, 2003.

22. Guo, L.; Zhu, S.; Zhang, S.; He, Q.; Li, W. Theoretical Studies of Three Triazole Derivatives as Corrosion İnhibitors for Mild Steel in Acidic Medium. *Corros. Sci.* **2014,** *87,* 366–375.

23. Sun, H.; Ren, P.; Fried, J. R. The Compass Force Field: Parameterization and Validation for Phosphazenes. *Comput. Theor. Polym. Sci.* **1998,** *8,* 229–246.

24. Guo, L.; Safi, Z. S.; Kaya, S.; Shi, W.; Tüzün, B.; Altunay, N.; Kaya, C. Anticorrosive Effects of Some Thiophene Derivatives Against the Corrosion of Iron: A Computational Study. *Front. Chem.* **2018,** *6,* 155–168.

25. Geerlings, P.; De Proft, F. Conceptual DFT: The Chemical Relevance of Higher Response Functions. *Phys. Chem. Chem. Phys.* **2008,** *10,* 3028–3042.

26. Kaya, S.; Kaya, C. A New Method for Calculation of Molecular Hardness: A Theoretical Study. *Comput. Theror. Chem.* **2015,** *1060,* 66–70.

27. Parr, R. G.; Pearson, R. G. Absolute Hardness: Companion Parameter to Absolute Electronegativity. *J. Am. Chem. Soc.* **1983,** *105,* 7512–7516.

28. Kaya, S.; Kaya, C. A New Equation Based on İonization Energies and Electron Affinities of Atoms for Calculating of Group Electronegativity. *Comput. Theor. Chem.* **2015,** *1052,* 42–46.

29. Kaya, S.; Kaya, C. A New Equation for Calculation of Chemical Hardness of Groups and Molecules. *Mol. Phys.* **2015,** *113,* 1311–1319.

30. Roy, R. K.; Krishnamurti, S.; Geerlings, P. Local Softness and Hardness Based Reactivity Descriptors for Predicting İntra-and İntermolecular Reactivity Sequences: Carbonyl Compounds. *J. Phys. Chem. A.* **1998,** *102,* 3746–3755.

31. Bellafont, N. P.; Illas, F.; Bagus, P. S. Validation of Koopmans' Theorem for Density Functional Theory Binding Energies. *Phys. Chem. Chem. Phys.* **2015,** *17,* 4015–4019.

32. Parr, R. G.; Szentpaly, L.; Liu, S. Electrophilicity İndex. *J. Am. Chem. Soc.* **1999,** *121,* 1922–1924.

33. Kaya, S.; Kaya, C.; Guo, L.; Kandemirli, F.; Tüzün, B.; Uğurlu, İ.; Madkour, L. H.; Saraçoğlu, M. Quantum Chemical and Molecular Dynamics Simulation Studies on İnhibition Performances of Some Thiazole and Thiadiazole Derivatives Against Corrosion of İron. *J. Mol. Liq.* **2016,** *219,* 497–504.

34. Mızrak, B.; Orman, E. B.; Abdurrahmanoğlu, Ş.; Özkaya, A. R. Synthesis, Characterization and Electrochemical Properties of Novel Pyridine Phthalocyanine Derivatives. *J. Porphyr. Phthalocyan.* **2018,** *22,* 1–7.

35. Yabaş, E.; Bağda, E.; Bağda, E. The Water Soluble Ball-Type Phthalocyanine as new Potential Anticancer Drugs. *Dyes Pigment.* **2015,** *120,* 220–227.
36. Kaya, S.; Banerjee, P.; Saha, S. K.; Tüzün, B.; Kaya, C. Theoretical Evaluation of some Benzotriazole and Phospono Derivatives as Aluminum Corrosion İnhibitors: DFT and Molecular Dynamics Simulation Approaches. *RSC Adv.* **2016,** *6,* 74550–74559.
37. Zuriaga-Monroy, C.; Oviedo-Roa, R.; Montiel-Sánchez, L. E.; Vega-Paz, A.; Marín-Cruz, J.; Martínez-Magadán, J. M. Theoretical Study of the Aliphatic-Chain Length's Electronic Effect on the Corrosion Inhibition Activity of Methylimidazole-Based Ionic Liquids. *Ind. Eng. Chem. Res.* **2016,** *55,* 3506–3516.
38. Erdoğan, S.; Safi, Z. S.; Kaya, S.; Özbakır Işın, D.; Guo, L.; Kaya, C. A Computational Study on Corrosion İnhibition Performances of Novel Quinoline Derivatives Against the Corrosion of İron. *J. Mol. Struct.* **2017,** *1134,* 751–761.
39. Oguzie, E. E.; Li, Y.; Wang, S. G.; Wang, F. Understanding Corrosion İnhibition Mechanisms-Experimental and Theoretical Approach. *RSC Adv.* **2011,** *1,* 866–873.
40. Singh, P.; Ebenso, E. E.; Olasunkanmi, L. O.; Obot, I. B.; Quraishi, M. A. Electrochemical, Theoretical, and Surface Morphological Studies of Corrosion Inhibition Effect of Green Naphthyridine Derivatives on Mild Steel in Hydrochloric Acid. *J. Phys. Chem. C.* **2016,** *120,* 3408–3419.
41. Obot, I. B.; Obi-Egbedi, N. O. Theoretical Study of Benzimidazole and its Derivatives and their Potential Activity as Corrosion İnhibitors. *Corros. Sci.* **2010,** *52,* 657–660.
42. Pearson, R. G.; Songstad, J. Application of the Principle of Hard and Soft Acids and Bases to Organic Chemistry. *J. Am. Chem. Soc.* **1967,** *89,* 1827–1836.
43. Lukovits, I. Kalman, E. Zucchi, F. Corrosion İnhibitors: Correlation between Electronic Structure and Efficiency. *Corrosion.* **2001,** *57,* 3–8.
44. Kovacevic, N.; Kokalj, A. Analysis of Molecular Electronic Structure of İmidazole- and Benzimidazole-Based İnhibitors: a Simple Recipe for Qualitative Estimation of Chemical Hardness. *Corros. Sci.* **2011,** *53,* 909–921.
45. Khaled, K. F. Molecular Simulation, Quantum Chemical Calculations and Electrochemical Studies for İnhibition of Mild Steel by Triazoles. *Electrochim. Acta.* **2008,** *53,* 3484–3492.
46. Oguzie, E. E.; Oguzie, K. L.; Akalezi, C. O.; Udeze, I. O.; Ogbulie, J. N.; Njoku, V. O. Natural Products for Materials Protection: Corrosion and Microbial Growth İnhibition using Capsicum Frutescens Biomass Extracts. *ACS Sustain. Chem. Eng.* **2013,** *1,* 214–225.
47. Wang, Z.; Lv, Q.; Chen, S.; Li, C.; Sun, S.; Hu, S. Effect of İnterfacial Bonding on İnterphase Properties in SiO_2/epoxy Nanocomposite: A Molecular Dynamics Simulation Study. *ACS Appl. Mater. Interfaces.* **2016,** *8,* 7499–7508.

CHAPTER 7

FORMULATION OF CELLULOSE ACETATE MEMBRANES INCORPORATION WITH MARJORAM AND PELARGONIUM ESSENTIAL OILS: EVALUATION OF ANTIMICROBIAL AND ANTIOXIDANT ACTIVITIES

T. M. TAMER[1, 2], Z. G. XIAO[1,*], Q. Y. YANG[1,*],
P. WANG[1], A. S. M. SALEH[1,3], N. WANG[1], and L. YANG[1]

[1]College of Grain Science and Technology, Shenyang Normal University, Shenyang, Liaoning 110034, China

[2]Polymer Materials Research Department, Advanced Technologies and New Materials Research Institute (ATNMRI), City of Scientific Research and Technological Applications (SRTA-City), New Borg El-Arab City, 21934, Alexandria, Egypt

[3]Department of Food Science and Technology , Faculty of Agriculture, Assiut University, Assiut, Egypt

*Corresponding authors. E-mail: zhigang_xiao@126.com, yangqy0311@163.com

ABSTRACT

By definition, the package is a bundle of something to isolate and protect it from its surroundings; it has a passive action about food, merely acting as a shield between it and the surrounding environment. Nonetheless, the incorporation of bio-active materials could improve beneficial interactions with food, such as antimicrobial action. Among several bioactive materials, essential oils appear as a promising materials result of their natural origin gives a sense of security to consumer enhancement of natural based polymeric membranes for active packaging taken the attention of scientists. The current work aimed to produce bio-active films of cellulose acetate (CA) by incorporation with marjoram and pelargonium extract oil. The influence of addition of different amounts of essential oils on wettability and mechanical properties was studied. Antibacterial activity of membranes was against Bacillus subtilis and Escherichia coli; it was increased as marjoram, and pelargonium extract oil percentage increased in CA film. The essential oils cellulose membranes blend films showed higher activity against Bacillus subtilis than that against Escherichia coli. Moreover, free radical scavenger activity (2, 2'-azino-bis (3-ethylbenzothiazoline-6-sulphonic acid and 2,2-diphenyl-1-picrylhydrazyl) of CA films increased as essential oil content increased. The obtained results provide high potential for the production of biodegradable food packaging film from CA and marjoram and pelargonium extract oil.

7.1   INTRODUCTION

Food packaging can be described as a coordinated system of preparing foods or nutrients for transport, distribution, storage, retailing, and use of the goods. The function of packaging can be divided into technical functions (included; contain, protect, measure, dispense, preserve, and store food) and marketing functions (included; communicate, promote, display, inform, motivate, and sell foods),[1] whereas marketing professionals need artistic and motivational understanding, and technical packaging needs science and engineering skills. Packaging is not a recent appearance. It is a process closely connected with the society evaluation and can be traced back to human civilization. Nature, degree, and amount of packaging at any society's stage reflect the needs, cultural patterns, materials availability, and technology of those societies. An excessive interest in food packaging was associated with industrial revolutions. At this time, rural agriculture workers migrated into cities and employed in factories. The consumer society has appeared. Factory workers needed commodities and food that was previously produced largely at home. By necessity, some industries located in nonagricultural areas, requiring that all food be transport into.

Food packaging is an activity closely associated with the evaluation of society, culture, and growth of civilization developments. Different materials were used starting from the shell of a nut, naturally hollow piece of wood, animal skin, plant leaves in the ancient period to wood, glasses, metal cans, and recently bags based on polymeric materials.

Recently, several polymeric materials are utilized in food packaging industries which include low- and high-density polyethylene, polyethylene terephthalate, biaxially oriented polypropylene, copolymer polypropylene, poly(vinyl chloride), and ethylene vinyl alcohol. No doubt, this massive usage of plastics is driven by several benefits including convenience and economics, but the drawbacks are also becoming apparent. Plastics do not biodegrade; primarily because they are made of synthetic polymers and no microbe has yet evolved that can feed on them. Disposal of the millions of tons of plastic waste generated every year takes up huge areas in the form of landfills. Plastic polymers may not be toxic themselves, but the myriad of chemical monomers added to them for improving their properties can be released to the surroundings and contact materials over time or under conditions such as heat and exposure to sunlight or photo-degradation. An example is bisphenol A (or BPA) that is added as a plasticizer but banned for use in applications involving packaging or containers for infant food due to its potential toxic effects. Waste plastics can also attract and accumulate chemical poisons present in the environment such as water contaminated with DDT and PCB .

Over the last few years, the attention of degradable polymer has increased in order to minimize the accumulation of hazardous wastes in the environment.[2-5] Recently, the application of degradable polymers covered specialized niche markets[6-9] such as packaging for the fast food industry. Customers and food producers have more requirements for packaging materials for further improvement in food quality and safety while still offering acceptably prolonged product shelf lives.[10] As a result, there is continually an inclusive interest for novel smart or bioactive food packaging systems that can meet these requirements acceptably and cost-effectively.[11] The utilization of antimicrobial materials in the area of smart antimicrobial packaging materials have attracted the attention as a result of the possible influence of those materials on product safety and shelf life. The most commercial food grade antimicrobial agents that utilized inedible food or packaging coatings are: sorbic acid, propionic acid, potassium sorbate, benzoic acid, sodium benzoate, and citric acid, bacteriocins, such as nisin and pediocin; enzymes, such as peroxidase and lysozyme; and polysaccharides displaying natural antimicrobial properties, such as chitosan.[12–16]

Cellulose acetate (CA) is a biobased polymer prepared via chemical acetylation of cellulose. It can be fabricated included many forms like membranes and beads nanofiber at low temperatures.[17] The unique mechanical properties of CA enable it to be used in many applications such as water treatment and filtration,[18,19] fuel cell applications,[20–22] packaging,[23] and medical purposes.[24]

Marjoram and pelargonium are used as spices and condiments. The main ingredient of marjoram essential oil was carvacrol which represented at 81.5%.[25] The main components of pelargonium were citronellol (33.6%), geraniol (26.8%), linalool (10.5%), citronellyl formate (9.7%), and p-menthone (6.0%).[26] The dill and marjoram oil could be used in food manufacturing such as sausage, fish, and fishery products to increase the acceptability of these products in addition to their effect on the shelf life.[27] Also, these oils have strong antimicrobial properties toward fungal and bacterial populations. As reported by Vagi et al.,[28] the ethanolic extract of marjoram showed antimicrobial activity against foodborne strains of bacteria and fungi. In the same field, it is found that the pelargonium essential oil has a significant rate of antioxidant and antibacterial activities.[29]

This study aims to prepare CA membranes by incorporation of marjoram and pelargonium essential oils. As far as we know, this is the first report on the incorporation of marjoram and pelargonium essential oils to CA films. The physical and thermal properties, antimicrobial activities and antioxidant activity of the prepared CA membranes with and without marjoram and pelargonium essential oils were evaluated. These properties may promote the film to be used in the cheese packaging industry.

7.2   MATERIALS

CA (degree of acetylation 40%) was supplied by Sigma-Aldrich Chemie Gmbh (USA). Marjoram and pelargonium extract oil, acetic acid, sulfuric acid; phenolphthalein, acetone, and sodium hydroxide were of analytical grade and obtained from Sinopharm Chemical Reagent Co., Ltd. (Beijing, China). Folin–Ciocalteu reagent and ABTS) were purchased from Sigma Aldrich Co., Ltd. (USA).

7.3 METHODS

7.3.1 PREPARATION OF MEMBRANE

CA-based membranes were prepared using traditional casting evaporation method. Briefly, 0.5 g of CA was dispersed in 25 mL of acetone and stirred for 6 h. Extracted oil (marjoram and pelargonium) was added to the CA solutions at concentrations of 0%, 20%, 50%, and 80% v/w (based on CA weight) under stirring and the solutions were marked as CA/Mar1, CA/Mar2, CA/Mar3 according to marjoram extract oil percentage, and CA/Prg1, CA/Prg2, CA/Prg3 for pelargonium extract oil, respectively, in addition to neat CA membrane. The mixtures were kept under stirring for 30 min to obtain a homogenous solution. The membrane solution was cast onto a clean glass Petri dish and allowed to dry at room temperature for next 72 h. The dried membranes were separated from the Petri dish and rinsed with 20 mL of distilled water. Finally, the wet membranes were spread out and attached to the clean glass support with clamps and allowed to dry for 24 h at room temperature.

7.3.2 MEMBRANE CHARACTERIZATION

7.3.2.1 WATER UPTAKE

Water uptake (%) estimation was performed by placing a dried sample in distilled water for 120 min to reach the equilibrium swelling state; the membrane was filtered off, carefully bolted with a filter paper, and weighted. The water uptake was calculated by the following equation:

$$\text{Water uptake (\%)} = \frac{[M - M_o]}{M_o} \times 100,$$

where M is the weight of the swelled sample and M_o is the weight of the dry sample.

7.3.2.2   MECHANICAL PROPERTIES

Mechanical properties of CA membranes incorporated with varying amounts of extracted oil (marjoram or pelargonium) were investigated using a universal testing machine (AG-1S, SHIMADZU, Japan) according to ASTM D-882 standards for testing tensile properties of paper. These properties included the membrane thickness and maximum stress and strain to failure. Membrane thickness measurements were obtained with an electronic digital micrometer. All measurements were carried in triplicate.

7.3.2.3   TOTAL PHENOLIC CONTENT

The total phenolic content of samples was determined based on the reduction of Folin–Ciocalteu reagent from yellow to blue-colored compound.[30] Membrane sample of 50 mg was immersed in 5 mL ethanol to extract phenolic compounds. A volume of 0.5 mL membranes extract was added to 2.0 mL Folin–Ciocalteu reagent (10%, v/v) followed by addition of 2 mL 7.5% (w/v) sodium carbonate solution. The mixture was incubated at 50°C for 5 min before the absorption was measured at 760 nm using a spectrophotometer (UV-1200S, China). Gallic acid solutions (0–100 µg) were used to obtain the standard curve. The total phenolic content was expressed as microgram gallic acid equivalent per gram sample (µg GAE/g membrane).

7.3.3   ABTS ASSAY

7.3.3.1   ABTS RADICAL SCAVENGING ASSAY

For ABTS radical scavenging assay, the radical cations were preformed by the reaction of an aqueous solution of $K_2S_2O_8$ (3.30 mg) in water (5 mL) with ABTS (17.2 mg). The resulting bluish green radical cation solution was stored overnight in the dark below 0°C. A volume of 1 mL the solution was diluted to a final volume of 60 mL with distilled water. The sample extract was obtained as described in the determination of total phenolic content. A volume of 0.1 mL membrane extract was added to 2.0 mL of ABTS solution. The test tube was then incubated in the dark for 30 min at room temperature. The decrease in absorbance was measured at 734 nm using a UV–VIS spectrophotometer. The percentage inhibition of radicals was calculated using the following formula:

$$\text{Inhibition (\%)} = \left[\frac{\left(A_{control} - A_{sample}\right)}{A_{control}}\right] \times 100,$$

where $A_{control}$ is the absorbance of ABTS solution without extract and A_{sample} is the absorbance of the sample with ABTS solution.

7.3.3.2 2,2-DIPHENYL-1-PICRYLHYDRAZYL RADICAL SCAVENGING ACTIVITY

The antioxidant activity of the extract was measured with the 2,2-diphenyl-1-picrylhydrazyl (DPPH) method with slight modifications.[31] A solution of DPPH was freshly prepared by dissolving 6 mg DPPH in 50 mL methanol (about 0.3 mM). The sample extract was obtained as described in the determination of total phenolic content. The extract (2.5 mL) and DPPH solution (2.5 mL) were mixed in a test tube. The test tube was then incubated in the dark for 20 min at room temperature. The decrease in absorbance was measured at 517 nm using a UV–VIS spectrophotometer. The percentage inhibition of radicals was calculated using the following formula:

$$\text{Inhibition (\%)} = \left[\frac{\left(A_{control} - A_{sample}\right)}{A_{control}}\right] \times 100$$

where $A_{control}$ is the absorbance of the DPPH solution without extract and A_{sample} is the absorbance of the sample with the DPPH solution.

7.3.4 EVALUATION OF ANTIMICROBIAL ACTIVITY

7.3.4.1 BROTH EVALUATION METHOD

Antimicrobial activity of CA membranes incorporated with different amounts of extracted oil (marjoram or pelargonium) was measured according to a method reported earlier.[32] Briefly, the bacteria were incubated in Luria-Bertani medium (LB medium, 1% peptone, 0.5% meat extract, and 1% NaCl, pH = 7). The inoculation was conducted at 37°C for 24 h under shaking. The obtained bacterial suspension was diluted with the previous peptone medium solution. Then, 0.1 mL of diluted bacteria suspension was cultured in 10 mL liquid peptone medium, and 50 mg of membranes was loaded. The inoculated medium incubated at 37°C for 24 h. After incubation, the antibacterial activity was monitored by two methods; first by measuring the optical density of the culture medium at 620 nm and calculating the inhibition percentage using the following equation[33]:

$$\text{Inhibition percent (\%)} = \frac{[OD_b - OD_s]}{OD_b} \times 100,$$

where OD_b and OD_s are the optical density of bacterial culture with and without tested membranes.

In the second monitoring method, 50 µL of bacterial culture were spread on triplicate nutrient agar plates, which were incubated at 37°C for 24 h, and the numbers of colonies were counted. All measurements were carried in triplicate.

7.3.4.2 IN SITU EVALUATION METHOD

For our experiment, European light cheese slices obtained commercially were used for the analysis. The cheese slices were removed from their commercial packages and cut into certain dimensions (3 cm × 3 cm and thickness ≈ 1.5 mm) then packaged with CA membranes containing different amounts of cassia oil. The slices were sterilized by exposed to ultraviolet light for 15 min on both sides.[34] After that, they were experimentally contaminated on one side, by the spreading 50 µL of a cell suspension (10^6 CFU/mL) of B. subtilis or E. coli in the cheese surface. After drying the inoculum, the slices were placed on the films with the contaminated side in contact with it. The system was stored for 12 days at 4°C were analyzed at 0, 3, 6, 9, and 12 days.

At contact time the sample was immersed in diluted in 10 mL LB for 5 min with shaking after that 50 μL of bacterial culture were spread on triplicate nutrient agar plates, which were incubated at 37°C for 24 h, and the numbers of colonies were counted. All measurements were carried in triplicate.

7.3.4.3 STATISTICAL ANALYSIS

The data were analyzed by ANalysis Of VAriance (ANOVA) using the SPSS software (version 17.0, Statistical Package for the Social Sciences Inc., Chicago, USA) and Duncan's multiple range tests processed differences among mean values. Significance was defined at $p < 0.05$. All experiments were performed in triplicate.

7.4 RESULTS AND DISCUSSION

Water uptake of CA membranes and that incorporated with marjoram and pelargonium essential oils are shown in Figure 7.1. The figure demonstrates a gradual decrease in water content by increasing the oil content. The hydrophobic phenolic ingredients of marjoram and pelargonium essential oils can explain the behavior. The same results were observed with contact angle measurements (Table 7.1). CA has both hydrophilic groups (i.e., hydroxyl groups) and hydrophobic groups (i.e., acetate groups); ratio and distribution of both groups control the wettability behavior of membranes. Addition of oils extracts influences this balance toward the hydrophobic nature. The acetyl groups of CA matrix (hydrophobic region) interact with polyaromatic ingredients of the essential oil, and so the water uptake changed depending on this interaction. Those results agree with earlier work.[35-37] The limited water uptake of CA membranes that incorporation with oils was maximized its efficiency for food packaging applications.

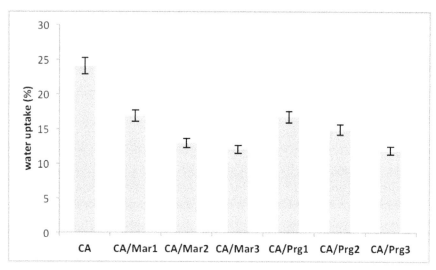

FIGURE 7.1   Water uptake of CA membranes incorporated with marjoram and pelargonium essential oils.

TABLE 7.1   Contact Angle Measurements of CA Membranes and That Incorporated with Different Content of Marjoram and Pelargonium Essential Oils.

	Contact angle (degree)
CA	53.7 ± 0.35
CA/Mar1	56.3 ± 0.17
CA/Mar2	71.8 ± 0.25
CA/Mar3	79.7 ± 0.33
CA/Prg1	62.5 ± 0.25
CA/Prg2	73.7 ± 0.32
CA/Prg3	76.9 ± 0.44

7.4.1   MECHANICAL PROPERTIES

Mechanical properties of CA membranes and that incorporated with marjoram or pelargonium oils are presented in Table 7.2. According to the table, the tensile strength (TS) of CA membrane was decreased with increasing oil content. For example, TS of CA (blank) was 151.5 N/mm², and it decreased to 124.7 and 68.5 N/mm² by using 0.5% of marjoram and pelargonium oils, respectively. The TS of CA membranes is attributed to the intermolecular hydrogen bonds between its functional groups (i.e., hydroxyl) along the polymer backbone. So the presence of essential oils causes the reduction in TS as it penetrates between the polymer chains and hinders the intermolecular interactions between CA functional groups.[38] Furthermore, presences of oils make change in CA membranes strain percent as this addition affects the arrangement of polymer chains.

TABLE 7.2 Mechanical Properties of CA Membranes Incorporated with Marjoram and Pelargonium Oil Extract.

Membranes	Mechanical properties	
	Stress (N/mm^2)	Strain (%)
CA	151.5 ± 7.2	15.3 ± 3.2
CA/Mar1	215.7 ± 11.3	13.3 ± 2.3
CA/Mar2	152.3 ± 9.1	14.9 ± 3.7
CA/Mar3	124.7 ± 7.3	21.7 ± 2.9
CA/Prg1	164.4 ± 8.7	26.5 ± 2.8
CA/Prg2	143.3 ± 9.2	33 ± 7.1
CA/Prg3	68.5 ± 5.3	14.7 ± 7.7

7.4.2 THERMAL PROPERTIES

Thermograms of CA membranes and that incorporated with marjoram and pelargonium oils membranes as obtained by DSC are shown in Figure 7.2. The DSC thermograms show an endothermic peak between 100°C and 150°C, which can be attributed to moisture evaporation.[39] On the other hand, exothermic peaks appeared at 250°C, which can be attributed to the decomposition of glucose pyranose ring along the polymer backbone.[40] The shift of peak to a higher temperature with the incorporation of essential oil indicates physical bond formatted between CA hydrophobic groups (acetate groups) and phenols and terpenoids in the essential oil.

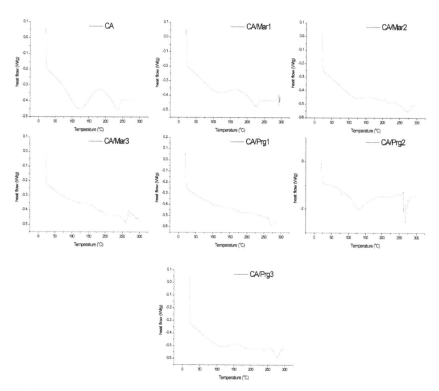

FIGURE 7.2 DSC thermogram of CA membranes incorporated with marjoram and pelargonium essential oils.

7.4.3 TOTAL PHENOLIC CONTENT

unique structure of phenolic compounds enables it to scavenge free radicals and perform as antioxidants.[41] Total phenolic compounds of extracts from membranes were estimated based on the reduction of Folin–Ciocalteu reagent, and the data are given in Figure 7.3. When the CA membranes were soaked in ethanol to extract included phenolic components, the membrane expended their structural integrity and liberated rosemary and aloe oil content. It can be seen that the total phenolic content of CA membrane increased after addition of marjoram and pelargonium essential oils. Moreover, it increased as the oil amount increased.

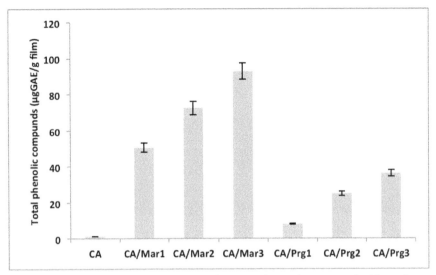

FIGURE 7.3   Total phenolic compounds of CA membranes incorporated with marjoram and pelargonium essential oils.

7.4.4   DPPH RADICAL SCAVENGING ACTIVITY DETERMINATION

In vitro design systems have been applied for evaluating the antioxidant potency of essential oil via measuring its free radical scavenging activity. DPPH assay based on 2,2-diphenyl-1-picrylhydrazyl radical for examining radical scavenging activity of the phenolic ingredients of essential oil or plant extracts. The radical form of DPPH has violet color until accepting an electron from the antioxidant compound, the violet color of the DPPH radical was reduced to yellow-colored diphenyl-picrylhydrazine radical that was estimated colorimetrically.[31]

Figure 7.4 presented the inhibition percent of DPPH dye by CA incorporated with the different percent of marjoram and pelargonium essential oils. The CA membrane has slightly radical scavenging activity due to the presence of polysaccharide hydroxyl group. The increase of marjoram and pelargonium essential oils in membranes shows a regular increase of free radical scavenging activity. The figure shows that antioxidant properties of marjoram extract are higher than that of pelargonium extract. This can be explained by the presence of different ingredients and phenolic compounds in both extracts. The obtained result is agreeing with that published in the literature.[42–44]

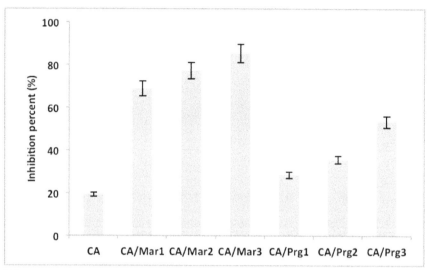

FIGURE 7.4   DPPH scavenging activity of CA membranes incorporated with marjoram and pelargonium essential oils.

7.4.5   ABTS RADICAL SCAVENGING ACTIVITY DETERMINATION

Antioxidant potential of CA membranes incorporated with marjoram and pelargonium essential oils was determined by ABTS[+]cation scavenging activity assay, and the results are shown in Figure 7.5.

From the Figure 7.5, it can be seen that CA membrane (control) showed the lowest ABTS radical scavenging activity. However, ABTS radical scavenging activity significantly increased after the addition of both marjoram and pelargonium essential oils. By increasing, the amount of oil from 20% to 80% (v/w) the activity was promoted. This can be attributed to the higher phenolic content of the oils compared with cellulose acetate. The presence of phenolic compounds in marjoram and pelargonium oils allowed it to donate an electron and decolorized ABTS[+] color.[42–44]

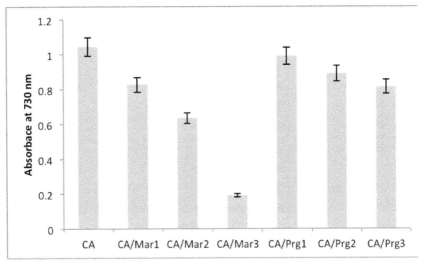

FIGURE 7.5   ABTS^{+}cation scavenging activity of CA membranes incorporated with marjoram and pelargonium essential oils.

7.4.6 ANTIBACTERIAL DETERMINATION

Evaluation of the antibacterial activity is essential to investigate the activity of membrane against pathogen bacterial for different applications. Figure 7.6 shows the antibacterial activity of the prepared membranes against *E. coli* (Gram-negative) and *B. subtilis* (Gram-positive) bacteria. The antibacterial activity was measured by growth turbidity method. The results showed that there is no antibacterial activity of cellulose acetate membrane. This result agrees with previous work.[45] On the other hand, the membranes acquire antimicrobial activity by adding marjoram and pelargonium essential oils to it. Furthermore, antimicrobial activity increased by increasing oil content. This result agrees with previous work.[45,46] The identification of natural antimicrobial compounds and the future development of these compounds through structure/activity studies provide a promising avenue of research for novel antimicrobials.

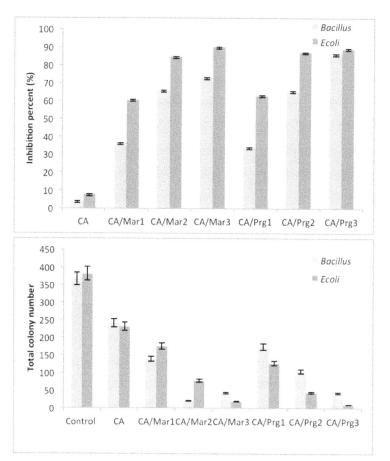

FIGURE 7.6 Antibacterial activity of CA membranes incorporated with marjoram and pelargonium essential oils against *Escherichia coli* and *Bacillus subtilis.*

The in situ investigation (sliced cheese) showed that the membranes with CA incorporated with marjoram and pelargonium extract oil were efficient in all concentrations tested. The counts of *B. subtilis* and *E. coli* were significantly reduced in cheese slices (Fig. 7.7). Untreated cheese slices show the high colony number of *B. subtilis* and *E. coli.* The same behavior was observed for treatment with CA membranes without any content of essential oils. It did not influence the development of the pathogen, with values equal to the untreated samples. This number significantly decreases by incorporation of essential oil pelargonium and much more in case of marjoram.

The significant decrease of development of bacterial was attributed to the antimicrobial activity of marjoram and pelargonium oil. In addition, the long-term control of antibacterial inhibition refers to the bactericidal activity of oil, in addition, the long-term oil release from the membrane structure.

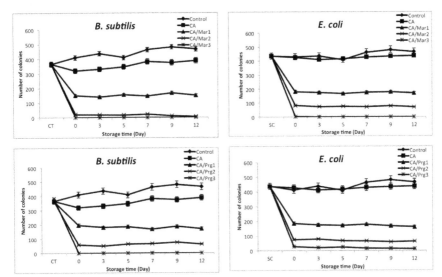

FIGURE 7.7 Antibacterial activity of CA membranes incorporated with marjoram and pelargonium essential oils against *Escherichia coli* and *Bacillus subtilis* in sliced cheese. CT, control treatment (without membrane).

7.5   CONCLUSION

The antibacterial packaging membranes based on CA were prepared through incorporation with marjoram and pelargonium oils.

The obtained results can be summarized as follows:

- The hydrophilic/hydrophobic character of CA membrane was influenced; incorporation of marjoram and pelargonium oil.
- Decreased the hydrophilicity of membranes.
- Incorporation enhanced the thermal stability of membranes with oil. DSC shows schiff of moisture band result elevation of marjoram and pelargonium oil at a higher temperature.

- Total phenolic compounds of membranes were increased by increasing the essential oil content especially for marjoram than pelargonium.
- Antioxidant evaluation is showing that the cellulose acetate/marjoram membrane has a higher tendency to scavenge free radicals than cellulose acetate/pelargonium membrane.
- The antibacterial evaluation demonstrates that this new membrane exhibited better antibacterial activity than neat CA membranes
- In experimentally contaminated cheese, the films were shown a gradual active against tested bacteria according to oil content percent.

The antimicrobial activity observed allows us to make a preliminary inference that CA films with marjoram or pelargonium are potential components for the production of active packaging. However for their effective use, toxicological evaluations should be performed to confirm the safety previously attributed to them.

KEYWORDS

- **cellulose acetate**
- **essential oil**
- **marjoram**
- **pelargonium**
- **antibacterial**
- **antioxidant**

REFERENCES

1. Coles, R.; McDowell, D.; Kirwan, M. J. Introduction. In *Food Packaging Technology*; Coles, R., McDowell, D., Kirwan, M. Eds.; CRC Press: Boca Raton, FL, 2003; pp 1–31.
2. Scott, G. Photo-biodegradable Plastics: Their Role in the Protection of the Environment. *Polym. Degrad. Stab.* **1990**, *29*, 135–154.
3. Evans, J. D.; Sikdar, S. K. Biodegradable Plastics: An Idea Whose Time Has Come? *Chemtech* **1990**, *20* (1), 38–42.
4. Nir, M. M.; Milty, J.; Ram, A. *Plastics Eng.* **1993**, *393*, 75–93.

5. Huang, S. J.; J. M. S. *Pure Appl. Chem.* **1995**, *A32* (A), 593–597.

6. Krupp, L. R.; Jewell, W. J. Biodegradability of Modified Plastic Films in Controlled Biological Environments. *Environ. Sci. Technol.* **1992**, *26*, 193–198.

7. Wailand, M.; Daro, A.; David, C. Biodegradation of Thermally Oxidized Polyethylene. *Polym. Degrad. Stab.* **1995**, *48*, 275–289.

8. Shah, P. B.; Bandopadhyay, S.; Bellare, J. R. Environmentally Degradable Starch Filled Low-density Polyethylene. *Polym. Degrad. Stab.* **1995**, *47*, 165–173.

9. Swift, J. *J.M.S. Pure Appl. Chem.* **1995**, *A32* (4), 641–651.

10. Theron, M. M.; Lues, J. F. R. Organic Acids and Meat Preservation: A Review. *Food Rev. Int.* **2007**, *23*, 141–158.

11. Cruz-Romero, M.; Kerry, J. The Packaging of Cooked Meats and Muscle-based Convenience-style Food Products. In *Processed Meats: 303 Improving Safety Nutrition and Quality*; Kerry, Joseph; Kerry, John, Eds.; Woodhead Publishing Limited: Cambridge, 2011; pp 666–705.

12. Quintavalla, S.; Vicini, L. Antimicrobial Food Packaging in the Meat Industry. *Meat Sci.* **2002**, *62*, 373–380.

13. Devlieghere, F.; Vermeulen, A.; Debevere, J. Chitosan: Antimicrobial Activity, Interactions with Food Components and Applicability as a Coating on Fruit and Vegetables. *Food Microbiol.* **2004**, *21*, 703–714.

14. Durango, A. M. Soares, N. F. F.; Andrade, N. J. Microbiological Evaluation of an Edible Antimicrobial Coating on Minimally Processed Carrots. *Food Control* **2006**, *17*, 336–341.

15. Chien, P. J.; Sheu, F.; Yang, F. H. Effects of Edible Chitosan Coating on Quality and Shelf Life of Sliced Mango Fruit. *J. Food Eng.* **2007**, *78*, 225–229.

16. Kim, K. W.; Min, B. J.; Kim, Y. T.; Kimmel, R. M.; Cooksey, K.; Park, S. I. Antimicrobial Activity Against Foodborne Pathogens of Chitosan Biopolymer Films of Different Molecular Weights. *LWT Food Sci. Technol.* **2011**, *44*, 565–569.

17. Gouvêa, D. M.; Mendonça, R. C. S.; Soto, M. L.; Cruz, R. S. Acetate Cellulose Film with Bacteriophages for Potential Antimicrobial use in Food Packaging. *LWT Food Sci. Tech.* **2015**, *63*, 85–91.

18. Mendes, G.; Faria, M.; Carvalho, A.; Gonçalves, M. C.; Pinho; M. N. Structure of Water in Hybrid Cellulose Acetate-Silica Ultrafiltration Membranes and Permeation Properties. *Carbohyd. Polym.* **2018**, *189*, 342–351.

19. El-Gendi, A.; Abdallah, H.; Amin, A.; Amin, S. K. Investigation of Polyvinyl Chloride and Cellulose Acetate Blend Membranes for Desalination. *J. Mol. Struct.* **2017**, *1146*, 14–22.

20. MohyEldin, M. S.; AbdElmageed, M. H.; Omer, A. M.; Tamer, T. M.; Youssef, M. E.; Khalifa, R. E. Novel Aminated Cellulose Acetate Membranes for Direct Methanol Fuel Cells (DMFCs). *Int. J. Electrochem Sci.* **2017**, *12*, 4301–4318.

21. MohyEldin, M. S.; AbdElmageed, M. H.; Omer, A. M.; Tamer, T. M.; Youssef, M. E.; Khalifa, R. E. Novel Proton Exchange Membranes Based on Sulfonated Cellulose Acetate for Fuel Cell Applications: Preparation and Characterization. *Int. J. Electrochem Sci.* **2016**, *11*, 10171–10150.

22. MohyEldin, M. S.; AbdElmageed, M. H.; Omer, A. M.; Tamer, T. M.; Youssef, M. E.; Khalifa, R. E. Development of Novel Phosphorylated Cellulose Acetate

Polyelectrolyte Membranes for Direct Methanol Fuel Cell Application. *Int. J. Electrochem. Sci.* **2016**, *11*, 3467–3491.

23. Duran, M.; Aday, M. S.; Zorba, N. N. D.; Temizkan, R.; Buyukcan, M. B.; Caner, C. Potential of Antimicrobial Active Packaging 'Containing Natamycin, Nisin, Pomegranate and Grape Seed Extract in Chitosan Coating' to Extend Shelf Life of Fresh Strawberry. *Food Bioprod. Process* **2016**, *98*, 354–363.

24. Gomaa, S. F.; Madkour, T. M.; Moghannem, S.; El-Sherbiny, I. M. New Polylactic Acid/cellulose Acetate-based Antimicrobial Interactive Single Dose Nanofibrous Wound Dressing Mats. *Int. J. Biol. Macromol.* **2017**, *105*, 1148–1160.

25. Daferera, D. J.; Ziogas, B. N.; Polissiou, M. G. The Effectiveness of Plant Essential Oils on the Growth of Botrytis Cinerea, Fusarium sp. and Clavibacter Michiganensis subsp. Michiganensis. *Crop Prot.* **2003**, *22*, 39–44.

26. Rana, V. S.; Juyal, J. P.; Blazquez, M. A. Chemical Constituents of Essential Oil of Pelargonium Graveolens Leaves. *Int. J. Aromather.* **2002**, *12* (4), 216–218.

27. Effat Afifi, A. A. Antimicrobial Potency of Some Natural Herbal Essential Oils. *Bull. Nutr. Inst. Cairo Egypt* **2001**, *21*, 1–15.

28. Vagi, E.; Simandi, B.; Suhajda, A.; Hehelyi, E. Essential Oil Composition and Antimicrobial Activity of *Origanum Majorana* L. Extract Obtained with Ethyl Alcohol and Supercritical Carbon Dioxide. *Food Res. Int.* **2005**, *38*, 51–57.

29. Ouedrhiri, W.; Balouiri, M.; Bouhdid, S.; Harki, E.-H.; Moja, S.; Greche, H. Antioxidant and Antibacterial Activities of Pelargonium Asperum and Ormenis Mixta Essential Oils and Their Synergistic Antibacterial Effect. *Environ. Sci. Pollut. Res.* **2017**. DOI: 10.1007/s11356-017-9739-1.

30. Navajas, Y. R.; Martos, M. V.; Sendra, E.; Alvarez, J. A. P.; López, F. In vitro Antibacterial and Antioxidant Properties of Chitosan Edible Films Incorporated with Thymus Moroderi or Thymus Piperella Essential Oils. *Food Control* **2013**, *30* (2), 386–392.

31. Dehpour, A. A.; Ebrahimzadeh, M. A.; Nabavi, S. F.; Nabavi, S. M. Antioxidant Activity of Methanol Extract of Ferula Assafoetida and Its Essential Oil Composition. *Grasas Y Aceites* **2009**, *60*, 405–412.

32. Skyttä, E.; Mattila, S. T. A Quantitative Method for Assessing Bacteriocins and Other Food Antimicrobials by Automated Turbidometry. *J. Microbiol. Meth.* **1991**, *14*, 77–88.

33. Vásquez, M. J. M.; Buitimea, E. L. V.; Jatomea, M. P.; Encinas, J. C. E.; Félix, F. R.; Valdes, S. S. Functionalization of Chitosan by a Free Radical Reaction: Characterization, Antioxidant and Antibacterial Potential. *Carbohydr. Polym.* **2017**, *155*, 117–127.

34. Lee, J. H.; Lee, J.; Bin, S. K. Development of a Chicken Feet Protein Film Containing Essential Oils. *Food Hydrocoll.* **2015**, *46*, 208–215.

35. Wu, J.; Chen, S.; Ge, S.; Miao, J.; Li, J.; Zhang, Q. Preparation, Properties and Antioxidant Activity of an Active Film from Silver Carp *(Hypophthalmichthys molitrix)* Skin Gelatin Incorporated with Green Tea Extract. *Food Hydrocoll.* **2013**, *32*, 42–51.

36. Park, S.; Zhao, Y. Incorporation of a High Concentration of Mineral or Vitamin into Chitosan-Based Films. *J. Agr. Food Chem.* **2004**, *52*, 1933–1939.

37. Peng, Y.; Wu, Y.; Li, Y. Development of Tea Extracts and Chitosan Composite Films for Active Packaging Materials. *Int. J. Biol. Macromol.* **2013**, *59*, 282–289.
38. Kalaycıoglu, Z.; Torlak, E.; Evingur, G. A.; Ozen, I.; Erim, F. B. Antimicrobial and Physical Properties of Chitosan Films Incorporated with Turmeric Extract. *Int. J. Biol. Macromol.* **2017**, *101*, 882–888.
39. Cheung, M. K.; Wan, K. P. Y.; Yu, P. H. Miscibility and Morphology of Chiral Semicrystalline Poly-(R)-(3-hydroxybutyrate)/Chitosan and Poly-(R)-(3-hydroxybutyrate-co-3-hydroxyvalerate)/Chitosan Blends Studied with DSC, 1H T1, and T1CRAMPS. *J. Appl. Polym. Sci.* **2002**, *86*, 1253–1258.
40. Kittur, F. S.; Prashanth, H.; Sankar, K. U.; Tharanathan, R. N. Characterization of Chitin, Chitosan and Their Carboxymethyl Derivatives by Differential Scanning Calorimetry. *Carbohydr. Polym.* **2002**, *49*, 185–193.
41. Lopez, M.; Martınez, F.; Del Valle, C.; Ferrit, M.; Luque, R. Study of Phenolic Compounds as Natural Antioxidants by a Fluorescence Method. *Talanta* **2003**, *60*, 609–616
42. Jhang, L. J.; Lee, D. J. Compressional-puffing Pretreatment for Enhanced Aantioxidant Compounds Extraction from *Aloe vera. J. Taiwan Inst. Chem. Eng.* **2017**, *81*, 170–174.
43. Vidic, D.; Tarić, E.; Alagić, J.; Maksimović, M. Determination of Total Phenolic Content and Antioxidant Activity of Ethanol Extracts from Aloe spp. *BCTBH* **2014**, *42*, 5–10.
44. Erkan, N.; Ayranci, G.; Ayranci, E. Antioxidant Activities of Rosemary (*Rosmarinus Officinalis* L.) Extract, Blackseed (*Nigella sativa* L.) Essential Oil, Carnosic Acid, Rosmarinic Acid, and Sesamol. *Food Chem.* **2008**, *110*, 76–82.
45. Liakos, L. L.; Holban, A. M.; Carzino, R.; Lauciello, S.; Grumezescu, A. M. Electrospun Fiber Pads of Cellulose Acetate and Essential Oils with Antimicrobial Activity. *Nanomaterials* **2017**, *7*, 84–93.
46. Ferro, V. A.; Bradbury, F.; Cameron, P.; Shakir, E.; Rahman, S. R.; Stimson, W. H. In Vitro Susceptibilities of Shigella flexneri and Streptococcus pyogenes to Inner Gel of Aloe Barbadensis Miller. *Antimicrob. Agents Chemother.* **2003**, *47*, 1137–1139.

CHAPTER 8

A PROMISING FUTURE OF CHEM DISCOVERY APPROACH: CHEMINFORMATICS EMERGING TECHNOLOGY

HERU SUSANTO[1,2*] and A. K. HAGHI[3]

[1]*Research Centre for Informatics, Indonesian Institute of Sciences, Cibinong, Indonesia*

[2]*Department of Computer Science and Information Management, Tunghai University, Taichung, Taiwan*

[3]*University Technology of Brunei, Bandar Seri Begawan, Brunei*

Corresponding author. E-mail: heru.susanto@lipi.go.id

ABSTRACT

Medical researchers are still struggling in the exploration of drugs for the treatment of diseases because of complicated arrangement of biomolecules that are accounted for infection disease, for instance, autism, AIDS, Alzheimer, cancer, and many more. Design and development for effective drug discovery, including information technology, is also used to predict and prevent the side effects caused by the newly discovered drug. The role of cheminformatics as an aid for the detection of drugs that are created in recent research of drug method has helped resolve the complicated matters associated with the old-fashioned system of drug discovery. Hence, cheminformatics assists therapeutic chemists for greater considerate of the complicated chemical compounds structure. The main purposes of using cheminformatics as a technology are gathering, keeping, and examining chemical data. This research emphasizes on cheminformatics in information system and information and communication technology, and their applications in drug discovery techniques, which are beneficial and useful for therapeutic researchers and chemists in discovering the solutions to complicated disease.

8.1 INTRODUCTION

Chemical informatics is a computer software system used to discover drugs and is applied to solve chemical problems. Cheminformatics system techniques are mainly used in pharmaceutical companies for drug processing. Prakash[55] stated the importance of cheminformatics software as a vital part of drug discovery that helps to solve chemical problems by storing drug details information, searching, and identifying the effects of drugs and retrieving and applying information about chemical compounds. He also mentioned cheminformatics applied in various aspect of scientific field involving chemistry, information, and computer science technology. Cheminformatics is clearly needed in the field of chemistry. Chemistry has been able to produce huge amount of data approximately around millions of chemical data each year. With cheminformatics tools and systems, with better management and proper databases, data information can be stored and securely accessed.

8.1.1 TRADITIONAL CHEMINFORMATICS VERSUS MODERN CHEMINFORMATICS

The traditional cheminformatics approach mentioned in Introduction to Drug Discovery (n.d.) use theoretical calculations. It is a complex situation and problem for scientists as an error or mistakes during evaluation and calculations are likely to occur throughout the process of drug discovery. The traditional discovery proved costly for researchers as it cost in excess of \$1.2–1.4 billion. Furthermore, it is time-consuming as it would take decades for the pharmaceutical industry to discover new drugs. During the discovery process, the drawbacks attributed by synthesizing compounds in the time-consuming multi-step process may lead to poor pharmacokinetics; drug clearance, in other words, removal of drugs from the body (39%); less effective (30%); animal safety test (11%); human side effects (10%); and other factors related in the cheminformatics.

Over the years, with rapid rise in competition, technology has made impacts and changed the culture of the pharmaceutical industry. Technology in the pharmaceutical industry has improved with a wide range of tools and software that can be applied in fields of drug discovery. Improved cheminformatics technology has a significant impact on processing and identifying drug-related problems and solutions.[63] Information technology has evolved over past decade as it helps to store vast quantities of chemical data, improve accuracy of chemical drug discovery, better and enhance decision making especially in an area of drug lead identification, and give best possible solutions and proper medications to reduce the drug effects. In addition, information technology including smart mobile devices emerging technology; helps modernize cheminformatics system to facilitate chemist to improve efficiency and productivity. Without cheminformatics, chemist must have to do everything manually; it would prove to be difficult for chemists and doctors as discovering a drug could be time-consuming and research development could be costly and moreover, it demands a huge amount of labor work, increasing paper work, and increasing cost, mistakes, and error might disrupt the workflow of the industry.[6,8,20,86,88]

Cheminformatics also requires molecular absorption, distribution, metabolism, excretion, and toxicity (ADMET), molecular modeling databases (MMDB), virtual screening and high-throughput screening system and quantitative structure activity relationship (QSAR) in aiding the drug discovery development process.

Another tool that is necessary for cheminformatics is known as high-throughput screening (HTS). It can take the advantage of automation by using a typically "friendlymechanism" multi-well that reads with a 96, 384, or 1536 well format to speedily determine the routine activity and behavior biological drug-like compounds, HTS, as well can also be effective for discovering ligands (a molecule that binds to another which is usually a larger molecule) for receptor–protein molecules that receive chemical signal from outside cells, enzymes, ion-chemicals, or other pharmacological targets. According to Hughes[42], HTS and compound screen are developed to detect the molecule compounds that interact with drugs.

In this chapter, we will explain more on the standard cheminformatic activities in the pharmaceutical industry like molecular modeling, virtual screening, HTS, large-scale data mining, and also to justify about the needs and importance of cheminformatics and lastly, its applications on drug discovery in the modern era.

8.2   LITERATURE REVIEW

8.2.1   MOLECULAR MODELING

The MMDB model that allows scientists and researchers to computer-based structure to visualize the molecules structure, predict, and simulating their behavior by using virtual screening and HTS to discover new lead compound for drugs or to refine drugs which then the data is stored through chemical libraries for future reference. In addition, it uses three-dimensional databases to visualize the molecule behavior. According to Chemica (2014), molecular modeling also known as computational chemistry aid in calculating and simulating the molecular routine behavior with specific equations. Chemica also added that molecular modeling are powerful computer systems to solve science-related problems. It represents one of the steps of drug discovery to find drug problems by utilizing model databases to solve complex problems in chemical aspects. Docking software is part of the MMDB according to Pagadala (2017). He also mentioned that it keeps close tabs on behavior of small molecules in identifies protein. The protein structure is experimented by using X-ray crystallography or nuclear magnetic resonance. Strategies of docking systems use calculations for lead optimizations. Molecular modeling is relatively cheap, convenient to use for data analyzing, interpreting data, identifying the molecule compound, and visualizing by using the three-dimensional structures.[5,12,18,79,85]

8.2.2   VIRTUAL SCREENING

Virtual screening is a technique that helps chemist to identify chemical structures by using in silico which has different technique such as bacterial sequencing techniques which is for identifying bacteria purposes, molecular modeling is used to demonstrate how drug interact with the nuclear cell receptors, whereas whole cell simulations is a technique that simulates the behavior of cell.[51] Virtual screening technique is a complementary technique to experimental HTS because according to Forli[37] virtual screening is much more "cheaper" than HTS which is defined as saving both money and the time as the process can be automated easily and also, although virtual screening does not help chemist professionals in terms of finding "The" active compound that going to provide the biological activities as it only will simulate what will happen in particular system through computer emulation but at least it helps them chemist professionals to reduce the number of compounds that was needed to test by giving the best-predicted compound and they will be able to prioritize to test the numbers of compound that most likely will bind with their target.

QSAR define by Dutta[34] structure–activity relations based using parameters to represent drug properties that have potential impact and influence on drug activities. In other words, it tracks and monitors the drug sequences using its 3D molecular modeling. Once 3D model is generated, docking systems will be further utilized to detect proteins–activity interaction. Lessigiarska[48] stated that using QSAR models are computer-based mathematical model that which relate the biological activity of compounds to their chemical structure. In addition, QSAR model helps better understanding of drug effects after being model. In QSAR analysis aspect, QSAR is used to describe relationship between chemical and biological activity.

Lessigiarska[48] pointed out the following purpose and functions of QSAR model use:

(1) With model used, it would give better understanding of interaction behavior between chemical and biological compounds.
(2) Information from the model can be used for experimental sample mainly in drug research and toxicity testing.
(3) With the model used, it saves time for producing large number of compound.

8.2.3 HIGH-THROUGHPUT SCREENING (HTS)

Another reason why cheminformatics is one of the needs and importance for chemist experts is because of its tool; HTS which is most likely to be used in order to help expertise to discover the new drug. As mentioned before, high throughput also can be effectively used to discover other pharmacological targets as, for example, to detect pathogens—a bacterium, virus, or other microorganisms that can cause disease. As mentioned in the book by Bhunia et al.,[29] pathogens can be found in contaminated food as well in the water and in the United States itself, pathogens have caused approximately around 48 million diseases, hospitalizations with 3000 death each year[56] and as 9 million causative agents that cause US foodborne illnesses are known but 38 million are not.[29] Hence it is needed for technologies to be developed to detect not only the known pathogen but also the unknown one, therefore, tools as HTS are used. The reason why is because according to Bhunia et al.,[29] HTS has the sensitivity platforms that are equipped with electronic sensors and specific biosensors for detection of even as small and light as *E. coli* bacterium, which are rod-shaped and are Gram-negative because they have a thin cell wall. By comparing traditional culture and based on non-culture with HTS, people can see how comfortable and less time consuming via HTS. On the other hand, traditional cultivation has a long process based on non-cultural development and also cannot detect some pathogens. This is due to a virus pathogens are difficult to detect through the methods available in this traditional development. However, this traditional approach is claimed to be more detailed and can keep the sample in a state of hygienist without pollutant contamination during the process. On the other hand, HTS offers a more precise method of approach to detect pathogens in moderate foods-mixed with microorganisms that are not targeted but with high sensitivity. Therefore it is very important to approach by combining the two methods above to produce more comprehensive and comprehensive findings.

8.2.4 LARGE-SCALE DATA MINING

Inside of every field, there are for sure data that are collected are kept in a clear place where it can be easily accessed when needed. It is kind of an urgent need for the new generation of computer technology to keep on evolved and assist human in terms simplifying the steps of retrieving information mainly from digital analysis. Therefore, this is where data mining can come to help in order to simplify the process. According to Jothi et al.[44] since the purpose of data mining is to help humanity in pattern discovery and extraction,[23] the process of data mining can include few procedures involving characterization, association, pattern correspond, meta-rule guided mining, and data screening image.

Jothi et al.[44] have said that data mining is the process of using technologies information software, involving uses of new technologies to store data. There are four disciplines that are considered as the data mining ground baseline which is machine learning,[65] artificial intelligence,[28] statistics,[46] and probability.[60] Kantardzic[45] said that data mining has two components of model which is predictive model, where its function is used to estimate the unfamiliar and unrecognized variable of interest. This model is frequently used in the healthcare and the other model of data mining is the descriptive model, which is mainly used for outlining the information about describing the data should be understandable by workers. Differences of the model can be differentiated by dividing the tasks. Descriptive models is determined by task such as task clustering,[27] association rule,[21] and also correlation analysis[40] while on the other hand, predictive models is determined by tasks such as task allocation and placement,[31] regression,[64] and also categorization (Genkan, Lewis, and Madigan, 2007). After knowing what are the models and also the task that is needed to be implemented, the next thing that is needed to know is the method of implementation to approach the discipline of data mining.

8.2.4.1 MOLECULAR MODELLING IN DATA MINING

"MMDB is anything that requires the use of a computer paint to describe or evaluate any aspects of the structure of a molecule" (Sanchez, n.d.). He also added that some methods are used in molecular modeling that include three-dimensional databases, analysis diversity, and ligands docking systems. After completion of molecular modeling, it can be stored in the databases for references. Keeping data secure and update is an important purpose in data mining. Compiling data is a procedure for data mining that creates and evaluates the model and lastly, use the knowledge as a platform.[1,17,19,68,87,91]

8.2.4.2 PROTEIN AND GENES

Gene articulation information cover-up imperative data required to comprehend the natural procedure that happens in a specific living being in connection to its condition, however, the multifaceted nature of organic systems and the volumes of qualities show increment the difficulties of appreciating and translation of the subsequent mass of information that comprises a huge number of estimation.[43] Thus, the utilization of grouping method is to address the test. This strategy has turned out to be helpful in making known the characteristic structure that is natural in quality articulation information, understanding quality capacity, cell forms, mining valuable data from uproarious information, and understanding quality control. Bunching technique additionally can recognize homology, which is essential for immunization plan. How does it make a difference? This technique (Bunching technique) Makes articulation of quality and comparison well-informed, to predict behavior and interaction between articulations. On the other hand Bunching Technique totally dislikes the information on the articulation of the quality of other groups disrupting the information produced by this articulation and comparison.

8.2.4.3   PREDICTION OF ANTI-CANCER ACTIVITIES

Dey et al.[59] point out that cancer problem is growing around the globe and it has increased by 33% in global cases between 2005 and 2015. The growth has been the highest in of low- and middle-income countries (LMICs) with the lowest development. There are only a few LIMCs that have control in preventing cancer tactics. Other countries have struggled in treating the cancer because of the expensive rate price and that would be a high ratio of most of the patient caught the symptoms of cancer at a later stage. Since most of LMICs are lacking sufficient treatment capability, the disease is growing drastically because the patient admits at later stages. In addition, a patient who suffers from cancer with not enough extensive care compounds is in all probability the diagnosis of cancer is death. The NCI Developmental Therapeutics Program (DTP) makes available dataset for 60 tumor cell line on forecast of cancer prevention activities which is dataset consider it is an important source for the expansion of recent agent of cancer prevention. Assay result is included in the dataset which quantifies the capacity of 44,653 compounds for the prevention of more than one or more cancer cell lines.

As Dey et al.[59] stated that the subset of the rest of the collection is made of estimation to 250,000 compounds and when screening data is vacant, the collection will denote the compounds. Many types of cancers (melanomas, leukemias, prostate breast cancers, and many more) have been accumulated more about 18 years. Hence, the statistics characterized dataset that contains harmful dose, development inhibition, and accumula-tion growth of inhibition. Moreover, toward the dosage interrelation data and cell lines that are not been treated already been are studied with the use of microarrays that provide gene expression. The article of the prediction of anticancer activities proves that it does need better high-performance equipment that can handle the large data mining in order for statistic or quality information for the cancer to be accurately executed.

8.4 RESULTS AND DISCUSSION

8.4.1 MOLECULAR MODELING

Molecular modeling is mentioned before by Chemica (2014); molecular modeling uses computer structures to create an image of the molecules structure to predict while monitoring the molecule activity with the aid of virtual screening and HTS. Furthermore, molecular model also helps to solve complex problems in discovery of new drugs.

Case study of elastin by Tarakanova and Buehler (2013) uses molecular modeling of protein materials. Proteins materials grow rapidly in area of research that contributes to fields of medicine and biology. They said that using advantage of molecular modeling as an experimental work in development and how molecular modeling has a key understanding in biomaterials.

Researchers point out that computational-based modeling has successful experiences in capturing data analysis in short period of time with better accurate and relevant information. Tarakanova and Bueher highlighted that there are two specific components required in molecular modeling which are systems' description and calculations scheme methods. Differences of two schemes are: systems' description involves in description and details of specific intra- and intermolecular interactions, whereas calculation scheme involves in numerical calculations with aim to reduce the potential risk of error.

Molecular modeling improves its reputations for which they have ability to connect the details of macroscopic and experiment of properties and mechanical compound. Elastin is a protein that protects structure in mammals found in tissues of human body. Moreover, elastin is produced in animal fetus. Damage or injury can have consequences which can be hard to replace. Emphysema and cutixlaxa are some diseases which are likely to affect the mammals. With the use of molecular modeling, it helps to exploit and use tissue engineering for development to counter the elastin effects.

8.4.2   VIRTUAL SCREENING

Virtual screening as stated by Vyas et al.[67] is cheminformatics tool used to identify new ligands, that is, molecules that bond to metal atom. Virtual screening also called *silico* screening has two kinds: structural-based screening "docking" and ligand-based virtual screening. Virtual screening represented a quick and financially savvy for databases screening in looking for drug leads. It is point out the differences in screening. However, structure-based screening is a tool used for design predict protein compound, whereas ligand based virtual screening focuses mainly on compound using others algorithm. The main aim of the virtual screening is to diminishing the gigantic virtual substance space for small organic molecules in order to produce or screen besides a specific target protein. So that it hinders the greatest opportunity to prompt a medication possibility for overseeing number of the compound. In theory, calculation using computer for finding properties of a compound is restricted by using virtual screening and problems arise while observing those properties. In contrast, when it comes to practical situation, the properties were calculated using database different from expectation. Numerous drug candidates were unsuccessful in the clinical trials due to mismatch of the effectiveness with the targeted drug, since software and hardware required yielding an appropriate answer.

Over than half of the clinical trials failed because of pharmacokinetic and toxicity issues. Hence, during the first phase of visual screening to calculate the resemblance of the drug of the same molecules is not aligning to their estimated drug target. Virtual screening has been utilized to define and evaluate large compound collections with the use of computer, so it could select the significant compounds for synthesis or assay. For that reason, the problem can be handled using extensive range of computational tactics. Vyas et al. concentrated on explicit receptor–ligand molecular docking where it could calculate the most precise model of the way so that a given ligand will tie to a receptor and as a result, candidates for synthesis or assay could utilize the most informative source for the assessment of the ligands. Even though virtual screening has been used with the basic principles in different ways for certain years, with the improved high-performance computing platforms, it does give satisfaction in noting the current impact on molecular modeling. The modeler can use affordable multiprocessor workstation computer units and personal computer clusters to engage computationally to execute algorithms on a routine basis. With the computationally intensive methods, this could give relevant date in the case of virtual screening, for instance, huge databases of chemical structure are for molecular docking appliance.

8.4.3 HIGH-THROUGHPUT SCREENING

Martis et al. (2010) mentioned that HTS is the technology used for discovering new drugs which is basically procedure of screening and assaying to enhance process of screening drug libraries that may usually exceed more than thousands per day.

Methods of HTS stated by Carnero[30] to identify new drugs proved to be difficult for scientist as they have to thoroughly investigate millions of compound libraries. HTS can collect huge amount of experiment sample in relatively shorter time. Carnero encourages and supports the implementation of HTS as it can record and test over thousands of compound within a day. With minimized time consumed, it can improve productivity and efficiency of the workers. HTS also requires larger chemical libraries since it can test more than thousand compounds per day.

As for case of HTS, one of the main benefits is to reduce time by testing huge amounts of samples in shorter time. HTS can be experiment to run test on toxicology. Toxicology as defined by Szymanski (2012) is science field that conducts research on growing substances in living organism. Szymanski mentioned that living organism is vulnerable to harmful substances that can cause severe damage the human body defenses and penetrate the bloodstream. Toxicology experiments are used for better understanding of chemical toxicity that has potential threats to human body systems. With improved technology in modern era, improve stem cell biology used as an opportunity for experimenting with the chemical compound. This experiment is quite useful and it is important as to collect as much information as possible to reduce the human safety risks environment. HTS also enhances assay process procedures as it relies on the use of machinery that can test more than thousands of compounds per day. To reduce the cost in animal use for experiment is crucial for HTS.

8.4.4   LARGE-SCALE DATA MINING

As for the case in healthcare, they are using data mining for purpose such as predicting the disease, and also, it acts as assistance for doctors in order to help them to do clinical decision. There are lots of methods that can be used in healthcare industries and few of them are anomaly detection, classification, clustering, and also decision tree. According to Luo et al. (2016), technologies of big data can collect large amount of biological clinical data at instant. With the applications of electronic health record which can process billions of DNA activities data. As the benefits of the extensive use of technology, it reduces the time for data analyse of biomedical processes, by developed of hardware and software through parallel computing and high performance computing processes.[7,13] Researchers categorizes data mining technologies into four phases:

(1) storing and extracting data;
(2) identifying the error and mistake;
(3) analyzing data;
(4) integrating platforms for developing solutions.

According to Luo et al. (2016), use of cloud burst helps to widen the data reading sequences with enhancing the speeding process 24 times faster than normal computer processor and identifying error to ensure that large-scale data can meet quality expectations. National Institutes of Health Cancer Genome created several tools, such as SAMQA, which is one of the tools used to identify error and find possible solutions to overcome the problems. This frameworks analysis and accommodate vast scale of DNA activity regularly. Finally, big data can benefit the researchers to adapt in different situations.

As stated in the research article of Jothi et al.[44] on the data mining in healthcare, few methods that can be used are anomaly detection, clustering, and decision tree. Anomaly detection is used in discovering significant change within data set.[36] Bio Lie et al. (2013) used three different anomaly detection method known as standard support vector data description, density-induced support vector data description, and also Gaussian mixture in order for him to check the accuracy of anomaly detection on uncertain dataset of liver disorder.[4]

Clustering on the other hand is usually used to seek to identify a finite set of categories or clusters to describe the data.[36] RuiVeloso (2014) used one of the clustering methods before which is known as vector quantization in predicting the readmission; a scene when a patient that has been discharged has the possibility to be admitted again within a specified time interval. Therefore, clustering can be such a handful aid for doctors to various types of patients who have the possibility to be admitted again after being discharged.

Decision tree is also one of the data mining methods that is commonly used. Decision support tools usually look like a tree-graph or model wherein it contain options of decision along with its consequences whenever a solution from the graph is chosen. Sharma and Om[58] and Wang et al.[61] used this method in their respective work in order for them to improve their performance in terms of accuracy.

8.4.4.1 MOLECULAR MODELING IN DATA MINING

In a case study of PubChem mentioned by Ming et al.,[53] PubChem was used for web search and data mining of natural products for their bioactivity. PubChem is a public database, containing vast amount of natural product. PubChem offers many tools to facilities search of chemical structure, bioactivity data, and molecular target information.

PubChem serves as a public repository for bioactivities through its own databases; BioAssay databases, PubChem gives an information system keep bioassay tracks records used for experiment chemical samples. PubChem provided web tools that have access from three databases to monitor the related bioactivity information. Wang[62] stated that PubChem collects valuable information from different kinds of sources that can be useful information to PubChem. They use information to data search, analysis, and retrieval tools to facilitate the use of structure and chemical results. Wang added that PubChem data use tools such as structure activity analysis to visualize and analyze the data.

8.4.4.2 PROTEIN AND GENES

Where designs as of now exist, examinations should likewise be possible to discover qualities whose articulation fits a particular wanted plan. Clustering could likewise be utilized to identify unidentified pathways to help handle illnesses. By grouping quality articulation information, qualities that are center casualties of assault of pathogens can be secluded, giving scientific experts a reasonable lead on medicate center. In 2000, Alizadeh et al.[25] utilized HC on DNA microarray information, and three unmistakable subtypes of the diffuse substantial B-cell lymphoma (DLBCL) were found. In 2015, lung growth datasets were examined to discover which kind of dataset and calculation would be best to analyze lung tumor. K-Means and farthest first calculations were utilized for the examinations. The K-Means calculation was observed to be productive for bunching the lung growth dataset with Attribute Relation File Format (ARFF).[33] Sirinukunwattana et al.[57] utilized the Gaussian Bayesian Hierarchical Clustering (GBHC) calculation. They tried the calculation of more than 11 malignancies and 3 engineered datasets. They understood that in contrast with other bunching calculations, the GBHC created more exact grouping which comes about medicinally affirmed. Karmilasari et al.[47] actualized K-implies calculation on pictures from the Mammography Image Analysis Society (MIAS) to decide the phase of dangerous bosom growth. Moore et al. (2010) distinguished five unmistakable clinical phenotypes of asthma utilizing unsupervised various leveled bunch investigation. All bunches contain subjects who meet the American Thoracic Society meaning of serious asthma, which bolsters clinical heterogeneity in asthma and the requirement for new methodologies for the order of sickness seriousness in asthma.

8.4.4.3 PREDICTION OF ANTICANCER CASE STUDY

As from the article about the prediction of anticancer as discussed by Dey et al.,[59] they also deliberated that Wang (2007) researched that they make an effort to accomplish large-scale data mining in excess of all dataset of 60 cell line using different kind of methods that comes from a SMARTS pattern abstraction numerical calculation to set choices of trees and unsystematic forests, Dey et al.[59] did their research that Breiman (1984) models were established and they have relation in their distributed infrastructure. To explain that the random forest models were picked because the certain favorable features, for instance, feature selection is not independent and to reduce data overload. They created 166 bit Molecular ACCess System (MACCS) keys to characterize the molecules. After that, they build the

A Random Forest framework for every cell line with the application in R server where the framework was created on a local machine; however, unable to connect into their platform. R server function is to detect available model files and transfers them into memory. Every model required estimated for about 16.5 min putting together and simple parallelization scheme is employed to give access to them to complete all of the models below than 8 h. Moreover, the data for every cell line is fairly unstable in terms of favor of inactive so that they use a biasing scheme which they quantified that to every tree in the collection that probably experimented differentially from the lively cell component that has been provided. Furthermore, the model performance was sensible even though the model is unstable as compared to all 60 models and it exhibiting about 67% of correct predictions overall and great model exhibiting about 77% correct prediction overall. The active class performance is measured a correct prediction which falls between 74% and 79% amongst all the 60 models. Once the frameworks developed, the researchers arrayed outside of R web service infrastructure. In addition, binary format is the place where all the 60 models were kept. It generates files of 3 models every time 20 models are being stored. R server is able to keep the files and it is basically needed for National Cancer Institute (NCI) random forest models; they required big space mainly because of memory usage. In consequences, it would take a lot of time to fill in. After it has been loaded into memory, it gives them an access by using Simple Object Access Protocol (SOAP)-based web service interface.

To be precise, they have developed a simple Java class that includes a Simplified Molecular Input Line Entry System (SMILES) string and it could automatically create the MACCS keys for the fill in the data molecule. After that, the class can be linked to R server through web service and gain a prediction. On the other side, a user can skip this class assistance and straight connect directly to R server via the public web service interface. In different occasion, they must need to input their own MACCS keys. Additionally, R server gives back the estimation of all the 60 cell lines alongside correct measurement (it is well-described in terms of the amount of the trees which assume the compound is lively). Lastly, usage data can be used in different matter. They also provide web page interface that can be easily make use of for the end-user and give access for the end-users to include SMILES string and gain different kind of textual and image illustrations of the results. From mentioned case study by Dey,

Dhillion, and Rajaman[59] that we can understand, it is important for the data mining being utilized to analyze such data and to develop awareness into the workings regarding chemical systems. According to the case study , the designed web service that depends on deployment platform that is appropriate for vast amount of chemical dataset is used to characterize an effort in modern drug discovery.

8.4.5 *MODERN DRUG DISCOVERY PROCESS*

Cheminformatics and technology systems integrated over the years with aims of achieving better productive and effective drug research and development, researches needs to take necessary steps by means of achieving better results and solutions for related problems. Cheminformatics tools and systems are all involved in drug discovery process. Begam and Kumar[26] stated that modern drug discovery involved with four vital processes:

(1) identification of target and validation by locating the source of the diseases,
(2) lead identification,
(3) lead optimization, and
(4) preclinical trials phase.

Identification target and validations first process of cheminformatics for drug discovery. Chen and Du[32] highlight that cheminformatics pinpoints the potential source of drug that causes diseases which could be either gene or protein. Cheminformatics use initial leads and scans computer databases or virtual libraries to select related compound. In addition, tools equipped by cheminformatics to ensure discovery drug development is completed and done electronically which includes design process with the use of computer-aided drug design. After identification target process completed, then the validation process begins to determine the drug outcome.[9,10,11]

Lead identification is a second-stage process. It helps to pinpoint and identify which chemical drug substances have a potential to treat disease. It uses HTS to ease the process of lead identifications. According to Maria and Anoop (n.d.), HTS is used to detect lead compound substances that likely have potential threats to disease. They added that HTS is a best possible tool as an alternative to screen massive libraries of chemical for the compound that focuses on the known structure. It can get to countless new ligands which can be purchased and tried. In addition, virtual screening offers a better handy course in finding new substances and lead to pharmaceutical research.[10,14]

Lead optimization utilizes distinctive systems and techniques that are utilized for lead identification. Methods used in lead optimizations consist of MMDB, QSAR, structure-based models, data mining, HTS, and ADMET

Preclinical trials are pivotal parts of drug discovery. It is a phase to identify whether the compound, which is safe and have minimal side effects, can be developed to cure a specific disease. Toxicity testing is performed to show how safe the drug is by the processes of absorption, distribution, metabolism, and excretion. In summary, preclinical trials are designed to achieve and avoid risk in development of new drugs.

ADMET, as mentioned by Ertl (n.d.), is a property that is related to the effects of the drugs and organism needed in pharmaceutical chemistry. However, he added that there are challenges and drawbacks of ADMET: it describes complex physical and biological processes and insufficient experimental data to build reliable models.

8.4.6 NEED AND IMPORTANCE OF CHEMINFORMATICS

Researches point out the effectiveness of cheminformatics technology which evolved over a decade. It represents huge step for pharmaceutical drug industry in near future. As mentioned before, it minimizes labor workforce and speeds up research discovery processing, minimizes sequences of discovery of new drugs. It shows a significant improvement from traditional drug discovery used over decade ago.

According to Begam and Kumar,[26] there are three main functions of cheminformatics:

(1) Developing and compiling data experiment and theory through information acquisition.
(2) Cheminformatics helps in information management that store data and retrieval information.

(3) Cheminformatics utilizes information knowledge that is applicable to data analysis, correlations, and solving problems in drug problems.

Drug discovery involved in lengthy time-consuming procedures including research and requires financial resources. Traditional drug discovery stated by Pharmacol (2011) that for researches to develop and discover new types of drugs can be costly which requires huge amount of money with approximately that access around more than $1 billion. He added that processing of new medications could be time consuming that takes a decade to finalize and launch it to the market. Traditional methods have a high chance of error and misleading information; some drugs may have side effects that can damage the body systems. To overcome the disadvantages of traditional drug discovery, this improved technology helps to minimize stress and burden for researches.[2,3]

In modern drug discovery, cheminformatics tools are equipped with high technology and advanced facilities that improve effectiveness and better discovery process. Bio- and cheminformatics devices are engaged with the early disclosure of lead possible candidate for a given target ailment. Cheminformatics is used in variety of applications. Molecular databases used in pharmaceutical drug industry are stated by Ertl (n.d.). The databases need to be checked for accuracy, and interface must be able to support data mining and can store vast amount of chemical data information. In addition, Ertl stated that with analyzing large quantity of data to obtain reliable information can lead to better understanding and made precise decisions. Cheminformatics uses technique of QSAR. QSAR is part of the MMDB techniques that uses numerical facts to locate biological structure activities.

QSAR is defined as tracking the activity of chemical compound, computing relevant molecular descriptor, assembling instructive models utilizing descriptors and applying the models. Models and databases might be visualized to help in model development and comprehension of compound libraries mentioned by Pirhadi et al. (2016). The main advantages of QSAR model as stated by Cherkasov (2014) is that QSAR contains useful and accurate information for drug discovery analysis. However, Cherkasov also pointed out the challenges and drawbacks of

QSAR models—insufficient data information can lead to incorrect data analysis which could affect the final results.

Computer analysis drug design (CADD) and computer-assisted synthesis design manufacturing (CASD) are technologies used extensively in pharmaceutical industry for drug development process. Begam and Kumar[26] highlighted that cheminformatics focuses on manipulate and analyses chemical model which can be modeled in two dimensions (2D) and three dimensions (3D). CADD and CASD provide tools for chemist to help analyze visually and can produce medication in short period of time. Fourches (n.d.) mentioned that CASD and CADD have significant role in cheminformatics. CASD and CADD computer applications are used in pharmaceutical drug industry as it creates, stores, and modifies model analysis electronically. In addition, CASD and CADD provide accurate and integrated information and produce output in short period of time.

CADD is an indispensable tool in the pharmaceutical industry. Kapenovic (2006) highlighted that CADD enhances the process of lead identification and lead optimizations. Moreover, it utilizes the uses of ADMET which prevent risk and error in drug design.

In addition, CADD has better understanding by suggesting which molecular structures to synthesize; it is cost effective for decision-making. Computer-aided designs have access of various software and resources in research and development field with main aim to discover along with optimizing the chemical compound. Hence, chemists and researchers should take accounts by utilizing of systems software and any resource related with CADD throughout their daily operations routine, to improve their skills and expertise with this technology implemented. On the other hand, CASD is a computer software system used to monitor and predict the chemical reactions and improve task designing model and analyzing model.

According to Tropsha and Bowen (1997), MMDB contains 3D structure base that consist of combinations of computerized chemistry and graphics and proved to be a valuable tool for drug research discovery. "With advanced software systems, it facilitate by providing better tools to integrate, explain and identify discovery of new phenomena" (Nadenda, 2004, p 51). Open Eye scientific software develops molecular modelling design since 1997 providing systems software for better use practices mainly in pharmaceutical industry. Areas of applications include cheminformatics drug discovery, docking systems, and lead optimizations for drug discovery.

As mentioned in this case study, the use of cheminformatics has many pharmaceutical benefits. Cheminformatics have large databases that can store large quantities of data. Case study of PubChem which they use such systems to store data. PubChem provides public databases. They used web-based design and data mining, whereas BioAssay databases are used to store large quantity of natural product and keep the chemical samples. PubChem also can access data from different sources which can be beneficial for further experiments and samples.

8.4.7 CHEMINFORMATICS AND ITS APPLICATIONS ON DRUG DISCOVERY

According to Aktar and Murmu,[24] Cheminformatics can be applied in some areas of the field in chemistry. The abundance applications of cheminformatics can be beneficial to any parts of chemistry. Such applications for chemical information includes in storing data created through experiment and can be extracted from chemical databases software libraries. In terms of analysis chemistry, information obtained from data collected can be utilized to predict on the source of desire objects. Cheminformatics are applicable for drug design process, identify lead, optimizing lead structure, and developing QSAR and ADMET. Applications with electronic and technology tools such as HTS that enable visual chemical compound for chosen activity. "Docking" systems and drug process are also part of the applied cheminformatics. Advantages of cheminformatics for pharmaceutical drug industry: it facilitates drug discovery process with the advanced tools and software technology such as data mining databases, virtual screening, and HTS and minimizes the labor workforce as it minimizes the use of paper work since it can be done electronically. Cheminformatics is a computerized structure-based systems such as molecular modeling that is used to visualize and analyze data in three-dimensional models.

Disadvantages of cheminformatics: it requires knowledge and skills; needs training for those workers who are computer illiterate; and could be time consuming. Implementing cheminformatics technology could be costly where the industry has to spend more than $1 billion and for the industry needs to hire high-skilled workers that are specialized in information system which may demand higher pay salary. However, for the industry that wants to improve its reputations by maintaining the high standards, it is worth a risk to implement such software systems. In addition, new technology might be difficult for senior workers as they need time to adapt and adjust in different environment and situations.

8.5 CONCLUSION

The combinations of chemical information systems and technology have changed the drug industry in modern times. Cheminformatics helps drugs industry by identifying the source that could be potential threat to the disease; cheminformatics tools help to reduce and minimize the labor workforce and speedup the process of discovery of new drugs. With proper databases, chemist library helps to store large amount of chemical data. With evolving technology, it modernizes the drug discovery process that helps to overcome the drawbacks of traditional drug discovery. It would take huge amount of labor workforce and research could cost more than 1 billion as mentioned before. As for discovery research process, it may take more than a decade to discover new types of drugs. In addition, chemistry can be quite complex to solved by method based on calculations. Virtual screening and HTS stated before has become an invaluable tools and play a significant role to identify chemical structure; also virtual screening is relatively cheap. Not to mentioned, it benefits the chemist and researchers itself as more knowledge instill on how to use the cheminformatics systems more effectively; with advanced cheminformatics technology, research can adapt on different situation and environment. It can help drug industry to improve effectiveness and efficiency among the workers to perform better with improve productivity and accuracy. It might be expensive to implement cheminformatics, depending on how willing the industry would take the risk to commit financially. Data mining of PubChem use large data mining to enable store large chemical data.

KEYWORDS

- chemistry
- chemical informatics
- chemical data
- information technology
- emerging technology

REFERENCES

1. Susanto, H.; Almunawar, M. N. Information Security Management Systems: *A Novel Framework and Software as a Tool for Compliance with Information Security Standard*; CRC Press, USA, 2008

2. Susanto, H.; Chen, C. K. Macromolecules Visualization Through Bioinformatics: An Emerging Tool of Informatics. *Appl. Phys. Chem. Multidisc. Appr.* **2018,** 383.

3. Susanto, H.; Chen, C. K. Informatics Approach and Its Impact for Bioscience: Making Sense of Innovation. *Appl. Phys. Chem. Multidisc. Appr.* **2018,** 407.

4. Susanto, H. Smart Mobile Device Emerging Technologies: An Enabler to Health Monitoring System. *Kalman Filtering Tech. Radar Track.* **2018,** 241.

5. Liu, J. C.; Leu, F. Y.; Lin, G. L.; Susanto, H. An MFCC□based text□independent Speaker Identification System for Access Control. *Concurr. Comput. Pract. Exp.* **2018,** *30* (2), e4255.

6. Almunawar, M. N.; Anshari, M.; Susanto, H.; Chen, C. K. How People Choose and Use Their Smartphones. In *Management Strategies and Technology Fluidity in the Asian Business Sector*; IGI Global, 2018; pp 235–252.

7. Susanto, H.; Chen, C. K.; Almunawar, M. N. Revealing Big Data Emerging Technology as Enabler of LMS Technologies Transferability. In *Internet of Things and Big Data Analytics Toward Next-Generation Intelligence*; Springer, Cham, 2018; pp 123–145.

8. Almunawar, M. N.; Anshari, M.; Susanto, H. Adopting Open Source Software in Smartphone Manufacturers' Open Innovation Strategy. In *Encyclopedia of Information Science and Technology*, 4th ed.; IGI Global, USA, 2018; pp 7369–7381.

9. Susanto, H. Cheminformatics: The Promising Future: Managing Change of Approach Through ICT Emerging Technology. *Appl. Chem. Chem. Eng. Princ. Methodol. Eval. Meth.* **2017,** *2*, 313.

10. Susanto, H. Biochemistry Apps as Enabler of Compound and DNA Computational: Next-Generation Computing Technology. *Appl. Chem. Chem. Eng. Exp. Tech. Method. Dev.* **2017,** *4*,181.

11. Susanto, H. Electronic Health System: Sensors Emerging and Intelligent Technology Approach. In *Smart Sensors Networks*, Elseviers B.V., The Netherland, 2017; pp 189–203.

12. Leu, F. Y.; Ko, C. Y.; Lin, Y. C.; Susanto, H.; Yu, H. C. Fall Detection and Motion Classification by Using Decision Tree on Mobile Phone. In *Smart Sensors Networks*, Elseviers B.V., The Netherland, 2017; pp 205–237.

13. Susanto, H.; Chen, C. K. Information and Communication Emerging Technology: Making Sense of Healthcare Innovation. In *Internet of Things and Big Data Technologies for Next Generation Healthcare*; Springer: Cham, 2017; pp 229–250.

14. Susanto, H.; Almunawar, M. N.; Leu, F. Y.; Chen, C. K. Android vs iOS or Others? SMD-OS Security Issues: Generation Y Perception. *Int. J. Technol. Diffus.* **2016,** *7* (2), 1–18.

15. Susanto, H. Managing the Role of IT and IS for Supporting Business Process Reengineering, 2016.

16. Susanto, H.; Kang, C.; Leu, F. Revealing the Role of ICT for Business Core Redesign, 2016.

17. Susanto, H.; Almunawar, M. N. Security and Privacy Issues in Cloud-Based E-Government. In *Cloud Computing Technologies for Connected Government*; IGI Global, USA, 2016; pp 292–321.

18. Leu, F. Y.; Liu, C. Y.; Liu, J. C.; Jiang, F. C.; Susanto, H. S-PMIPv6: An intra-LMA Model for IPv6 mobility. *J. Netw. Comput. Appl.* **2015,** *58,* 180–191.

19. Susanto, H.; Almunawar, M. N. Managing Compliance with an Information Security Management Standard. In *Encyclopedia of Information Science and Technology*, 3rd ed.; IGI Global, USA, 2015; pp 1452–1463.

20. Almunawar, M. N.; Susanto, H.; Anshari, M. The Impact of Open Source Software on Smartphones Industry. In *Encyclopedia of Information Science and Technology*, 3rd ed.; IGI Global, USA, 2015; pp 5767–5776.

21. Almunawar, M. N.; Anshari, M.; Susanto, H. Crafting Strategies for Sustainability: How Travel Agents Should React in Facing a Disintermediation. *Oper. Res.* **2013,** *13* (3), 317–342.

22. Almunawar, M. N.; Susanto, H.; Anshari, M. A Cultural Transferability on IT Business Application: iReservation System. *J. Hosp. Tour. Technol.* **2013,** *4* (2), 155–176.

23. Agrawal, R.; Psaila, G. Active Data Mining. *Current* **1995,** 3–8.

24. Aktar, M. W.; Murmu, S. Cheminformatics: Principle and Applications, 2018. http:// www.shamskm.com/files/chemoinformatics-principles-and-applications.pdf. (accessed May 14, 2019)

25. Alizadeh, A. A.; Eisen, M. B.; Davis, R. E., et al. Distinct Types of Diffuse Large B-cell Lymphoma Identified by Gene Expression Profiling. *Nature* **2004,** *403* (6769), 503–511.

26. Begam, B. F.; Kumar, J. S. A Study on Cheminformatics and Its Applications on Modern Drug Discovery. *Proc. Eng.* **2012,** 1264–1275.

27. Berkhin, P. A Survey of Clustering Data Mining Techniques. In *Grouping Multidimensional Data*; Springer, Berlin, Heidelberg, 2006, pp 25–71.

28. Bhattacharyya, D. K.; Hazarika, S. M. *Networks, Data Mining and Artificial Intelligence: Trends and Future Directions*, 1st ed.; Narosa Pub House, New Delhi, 2006.

29. Bhunia, A.; Kim, M.; Taitt, C. *High Throughput Screening for Food Safety Assessment*; Woodhead Publishing: United Kingdom, 2015.

30. Carnero, A. High Throughput Screening in Drug Discovery. *Clin. Trans. Oncol.* **2006,** *8* (7), 482–90.

31. Chau, R. M.; Cheng, R.; Kao, B.; Ng, J. Uncertain Data Mining: An Example in Clustering Location Data. *Lect. Notes Comput. Sci.* **2016,** *3918*, 199–204.

32. Chen, X. P.; Du, G. H. Target Validation: A Door to Drug Discovery. *Drug Discov. Ther.* **2007,** *1* (1), 23–29.

33. Dharmarajan, A.; Velmurugan, T. Lung Cancer Data Analysis by k-means and Farthest First Clustering Algorithms. *Ind. J. Sci. Technol.* **2015,** *8* (15).

34. Dutta, S. Application of QSAR in Drug Design and Drug Discovery. *Int. J. Recent Sci. Res.* **2015,** *6* (5), 3921–3924.

35. *E. coli and Food Safety Centers for Disease Control and Prevention*, 2017. https://www.cdc.gov/features/ecoliinfection/index.html (accessed Aug 31, 2017).

36. Fayyad, U.; Piatetsky-Shapiro, G.; Smyth, P. From Data Mining to Knowledge Discovery in Databases *AI Mag.* **1996,** 37–54.

37. Forli, S. *Introduction to Virtual Screening: Stefano Forli.* [Video], 2015. https://www.youtube.com/watch?v=8Q0tk6HtRuc. (accessed May 14, 2019).

38. Fourchs D.. Cheminformatics: At the Crossroad of Eras. In *Application of Computational Techniques in Pharmacy and Medicine. Challenges and Advances in Computational Chemistry and Physics*; Gorb, L., Kuz'min, V., Muratov, E., Eds; Springer: Dordrecht, 2014; Vol. 17.

39. Genkin, A.; Lewis, D. D.; Madigan, D. Large-Scale Bayesian Logistic Regression for Text Categorization. *Technometrics* **2007,** *49* (3), 291–304.

40. Hardoon, D. R.; Sandor, R. S.; John, R. S. Canonical Correlation Analysis: An Overview with Application to Learning Methods. *J. Neural Comput.* **2004,** *16* (12), 2639–2664.

41. Hong, T. P.; Lin, K. Y.; Wang, S. L.. Fuzzy Data Mining for Interesting Generalized Association Rules. *Fuzzy Sets Syst.* **2003,** *138* (2), 255–269.

42. Hughes, J. P. Principle of Early Drug Discovery. *Br. J. Phamacol.* **2011,** *162* (6), 1239–1249. DOI: https://dx.doi.org/10.1111%2Fj.1476-5381.2010.01127.x10.1111/j.1476-5381.2010.01127.x.

43. Jelili Oyelade, E. *Clustering Algorithms: Their Application to Gene Expression Data.PubMed Central (PMC)*, 2016. https://www.ncbi.nlm.nih.gov/pmc/articles/PMC5135122/ (accessed Sept 2, 2017).

44. Jothi, N.; Rashid, N.; Husain, W. Data Mining in Healthcare: A Review. *Proc. Comput. Sci.* **2015,** *72*, 306–313. http://dx.doi.org/10.1016/j.procs.2015.12.145. (accessed May 14, 2019).

45. Kantardzic, M. *Data Mining: Concept, Models, Methods, and Algorithms*, 2nd ed.; Wiley-IEEE Press, Canada, 2011.

46. Karegar, M.; Isazadah, A.; Fartash, F.; Saderi, T.; Navin, A. H. *Data-Mining by Probability-Based Patterns*; Proceeding of 30th International Conference on Information Technology Interfaces – ITI 2008, Croatia, IEEE, 2006; pp 353–360.

47. Karmilasari, S.W.; Hermita, M.; Agustiyani, N. P.; Hanum, Y.; Lussiana, E. T. P. Sample K-Means Clustering Method for Determining the Stage of Breast Cancer Malignancy Based on Cancer Size on Mammogram Image Basis. *Int. J. Adv. Comput. Sci. Appl.* **2015,** *5* (3), 86–90.

48. Lessigiarska, I. Development of Structure–Activity Relationship For Pharmacotoxi-cologicalEndpoitns Relevant To European Union Legislation, 2006. http://publica-tions.jrc.ec.europa.eu/repository/handle/JRC37649. (accessed May 14, 2019)

49. Liao, C.; Sitzman, M.; Pugliese, A.; Nicklaus, M. C. Software and Resources for Computational Medicinal Chemistry, 2012. https://www.ncbi.nlm.nih.gov/pmc/articles/PMC3413324. (accessed May 14, 2019).

50. Liu, B.; Xiao, Y.; Cao, L.; Hao, Z.; Deng, F. SVDD-based Outlier Detection on Uncertain Data. *Knowl. Inf. Syst.* **2013,** *34* (3), 597–618.

51. Marshall, T. Differences Between In Vitro, In Vivo, and In Silico Studies (MPKB). [online] Mpkb.org. https://mpkb.org/home/patients/assessing_literature/in_vitro_studies. (accessed May 14, 2019).

52. Martins, E. A.; Radhakrishanan, R.; Badve, R. R. High-Throughput Screening: The Hits and Leads of Drug Discovery—An Overview. *J. Appl. Pharm. Sci.* **2010,** *1* (1), 02–10.

53. Ming, H.; Tiejun, C.; Yanli, W.; Stephen, B. H. Web Search and Data Mining of Natural Products and Their Bioactivities in PubChem, 2013. https://www.ncbi.nlm.nih.gov/pmc/articles/PMC3869387/.

54. Nadendla, R. M. *Molecular Modeling: A Powerful Tool for Drug Design and Molecular Docking*; Siddhartha College of Pharmaceutical Sciences, Springer India, 2004; pp 51–60.

55. Prakash, N.; Gareja, D. A. Cheminformatics. *J. Prot. Bioinform.* **2010,** *3* (8), 249–252. DOI:10.4172/jpb.1000147.

56. Scallan, E.; Hoekstra, R. M.; Angulo, F. J.; Tauxe, R.V.; Widdowson, M. A.; Roy, S. L., et al. Foodborne Illness Acquired in the United States Major Pathogens. *Emerg. Infect. Dis.* **2011,** *17* (1), 7–15.

57. Sirinukunwattana, K.; Savage, R. S.; Bari, M. F.; Snead, D. R. J.; Rajpoot, N. M. Bayesian Hierarchical Clustering for Studying Cancer Gene Expression Data with Unknown Statistics. *PLoS One* **2013,** *8* (10), e75748.

58. Sharma, N.; Om, H. Data Mining Models for Predicting Oral Cancer Survivability. *Netw. Model. Anal. Heal. Inf. Bioinf.* **2013,** *2* (4), pp 285–295.

59. Subhojit, Dey; Preet K., Dhillon; Preetha, Rajaraman Cancer Prevention in Low- and Middle-Income Countries. *J. Cancer Epidemiol.* **2017,** DOI:10.1155/2017/8312064.

60. Thomas, H.; Paul, L. *Statistics: Methods and Applications*, 1st ed.; StatSoft, Inc, USA, 2005.

61. Wang, K.-J; Makond, B.; Wang, K.-M. An Improved Survivability Prognosis of Breast Cancer by Using Sampling and Feature Selection Technique to Solve Imbalanced Patient Classification Data. *BMC Med. Inform. Decis. Mak.* **2013,** *13*, 124.

62. Willett, P. *A Bibliometric Analysis of Literature of Cheminformatics*; University of Sheffield: UK, 2007.

63. Wu, Z.; Li, C. *L0-Constrained Regression for Data Mining*; Proceeding of Pacific-Asia Conference on Knowledge Discovery and Data Mining; Springer, Berlin, Heidelberg, 2007; pp 981–988.

64. Witten, I. H.; Frank, E.; Hall, M. *Data Mining: Practical Machine Learning Tools and Techniques*; Google eBook, 2011.

65. Veloso, R.; Portela, F.; Santos, M. F.; Silva, Á.; Rua, F.; Abelha, A.; Machado, J. A Clustering Approach for Predicting Readmissions in Intensive Medicine. *Procedia Technol.* **2014,** *16,* 1307–1316.

66. Vyas, V.; Jain, A.; Gupta, A. *Virtual Screening: A Fast Tools for Drugs Design*; Nahata College of Pharmacy, Department of Medicinal Chemistry, Rajiv Gandhi Technical University, 2008; Vol. 76, pp 333–360. DOI:10.3797/scipharm.0803-03.

67. Susanto, H.; Almunawar, M. N. *Information Security Management Systems: A Novel Framework and Software as a Tool for Compliance with Information Security Standard*; CRC Press, USA, 2018.

68. Susanto, H.; Chen, C. K. Macromolecules Visualization Through Bioinformatics: An Emerging Tool of Informatics. *Appl. Phys. Chem. Multidisc. Appr.* **2018,** *383.*

69. Susanto, H.; Chen, C. K. Informatics Approach and its Impact for Bioscience: Making Sense of Innovation. *Appl. Phys. Chem. Multidisc. Appr.* **2018,** *407.*

70. Susanto, H. Smart Mobile Device Emerging Technologies: An Enabler to Health Monitoring System. In *High-Performance Materials and Engineered Chemistry*; Apple Academic Press, USA, 2018; pp 241–264. 72.

71. Liu, J. C.; Leu, F. Y.; Lin, G. L.; Susanto, H. An MFCC☐based Text☐independent Speaker Identification System for Access Control. *Concurr. Comput. Pract. Expe.* **2018,** *30* (2), e4255.

72. Almunawar, M. N.; Anshari, M.; Susanto, H.; Chen, C. K. How People Choose and Use Their Smartphones. In *Management Strategies and Technology Fluidity in the Asian Business Sector*; IGI Global, USA, 2018; pp 235–252.

73. Susanto, H.; Chen, C. K.; Almunawar, M. N. Revealing Big Data Emerging Technology as Enabler of LMS Technologies Transferability. In *Internet of Things and Big Data Analytics Toward Next-Generation Intelligence*; Springer: Cham, 2018; pp 123–145.

74. Almunawar, M. N.; Anshari, M.; Susanto, H. Adopting Open Source Software in Smartphone Manufacturers' Open Innovation Strategy. In *Encyclopedia of Information Science and Technology*, 4th ed.; IGI Global, USA, 2018; pp 7369–7381.

75. Susanto, H. Cheminformatics—The Promising Future: Managing Change of Approach Through ICT Emerging Technology. *Appl. Chem. Chem. Eng. Prin. Methodol. Eval. Methods,* **2017,** *4,* 313.

76. Susanto, H. Biochemistry Apps as Enabler of Compound and DNA Computational: Next-Generation Computing Technology. *Appl. Chem. Chem. Eng. Exp. Tech. Method. Dev.* **2017,** *4,* 181.

77. Susanto, H. Electronic Health System: Sensors Emerging and Intelligent Technology Approach. In *Smart Sensors Networks*, Elseviers B.V., The Netherland, 2017c; pp 189–203.

78. Leu, F. Y.; Ko, C. Y.; Lin, Y. C.; Susanto, H.; Yu, H. C. Fall Detection and Motion Classification by Using Decision Tree on Mobile Phone. In *Smart Sensors Networks*, Elseviers B.V., The Netherland, 2017; pp 205–237.

79. Susanto, H.; Chen, C. K. Information and Communication Emerging Technology: Making Sense of Healthcare Innovation. In *Internet of Things and Big Data Technologies for Next Generation Healthcare*; Springer: Cham, 2017; pp 229–250.

80. Susanto, H.: Almunawar, M. N.; Leu, F. Y.; Chen, C. K. Android vs iOS or Others? SMD-OS Security Issues: Generation Y Perception. *Int. J. Technol. Diffus.* **2016,** *7* (2), 1–18.

81. Susanto, H. *IT Emerging Technology to Support Organizational Reengineering;* SSRN, Elseviers, The Netherland, 2016.

82. Susanto, H.; Almunawar, M. N. Security and Privacy Issues in Cloud-Based E-Government. In *Cloud Computing Technologies for Connected Government;* IGI Global, USA, 2016; pp 292–321.

83. Leu, F. Y.; Liu, C. Y.; Liu, J. C.; Jiang, F. C.; Susanto, H. S-PMIPv6: An intra-LMA Model for IPv6 Mobility. *J. Netw. Comput. Appl.* **2015,** *58,* 180–191.

84. Almunawar, M. N.; Anshari, M.; Susanto, H.; Chen, C. K. Revealing Customer Behavior on Smartphones. *Int. J. Asian Bus. Inf. Manag.* **2015,** *6* (2), 33–49.

85. Susanto, H.; Almunawar, M. N. Managing Compliance with an Information Security Management Standard. In *Encyclopedia of Information Science and Technology,* 3rd ed.; IGI Global, USA, 2015; pp 1452–1463.

86. Almunawar, M. N.; Susanto, H.; Anshari, M. The Impact of Open Source Software on Smartphones Industry. In *Encyclopedia of Information Science and Technology,* 3rd ed.; IGI Global, USA, 2015; pp 5767–5776.

87. Almunawar, M. N.; Anshari, M.; Susanto, H. Crafting Strategies for Sustainability: How Travel Agents Should React in Facing a Disintermediation. *Oper. Res.* **2013,** *13* (3), 317–342.

88. Almunawar, M.; Susanto, H.; Anshari, M. A Cultural Transferability on IT Business Application: iReservation System. *J. Hos. Tour. Technol.* **2013,** *4* (2), 155–176.

89. Susanto, H.; Almunawar, M. N.; Tuan, Y. C. Information Security Management System Standards: A Comparative Study of the Big Five. *Int. J. Electr. Comput. Sci.* **2011,** *11* (5), 23–29.

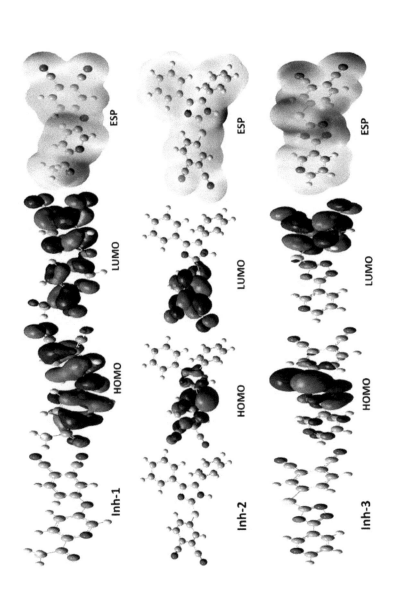

FIGURE 6.1 The optimized structures, HOMOs, LUMOs, and electrostatic potential structures of non-protonated inhibitor molecules using DFT/6-311++G** calculation level.

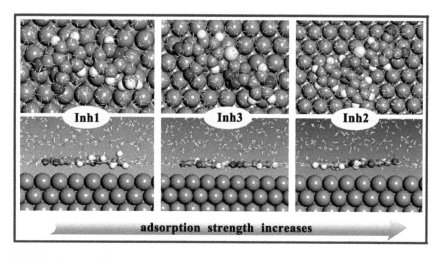

FIGURE 6.2   Equilibrium adsorption configurations of inhibitors *Inh-1*, *Inh-2*, and *Inh-3* on Fe(110) surface obtained by MD simulations. Upper: top view and lower: side view.

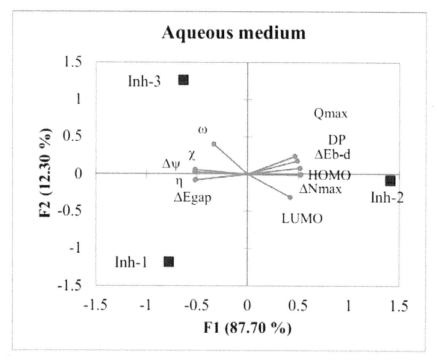

FIGURE 6.3   PCA based on a correlation matrix combining quantum chemical descriptors (red circles) and the studied inhibitors (*Inh-1*, *Inh-2*, and *Inh-3*; blue squares).

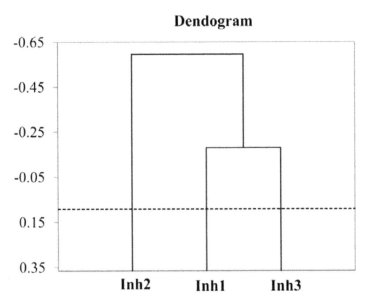

FIGURE 6.4 Dendrogram obtained for the studied inhibitors (*Inh-1*, *Inh-2*, and *Inh-3*) in aqueous phase.

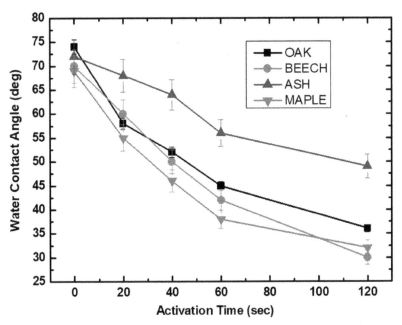

FIGURE 15.2 Water contact angle of RFD plasma-treated wood species vs. plasma activation time.

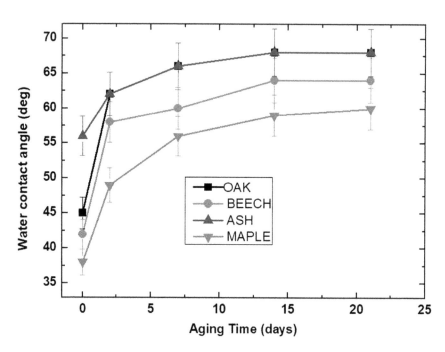

FIGURE 15.3   Water contact angle of RFD plasma-treated wood species vs. plasma activation time.

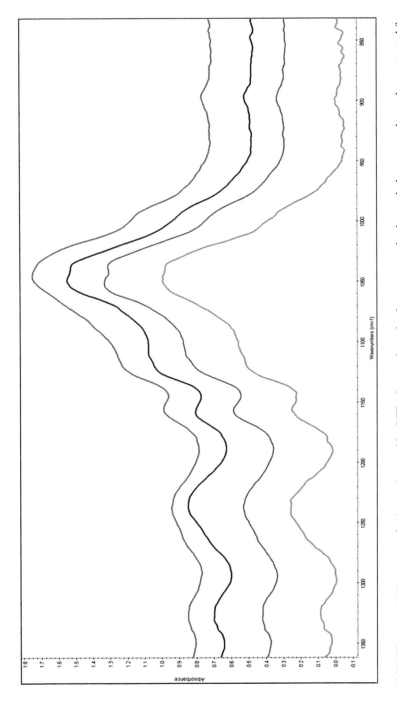

FIGURE 15.4 FTIR spectra of oak wood treated by RFD plasma: the red color—untreated oak wood, the green color—plasma-treated for 20 s, the blue color—plasma-treated for 60 s, and the purple color—plasma-treated for 120 s.

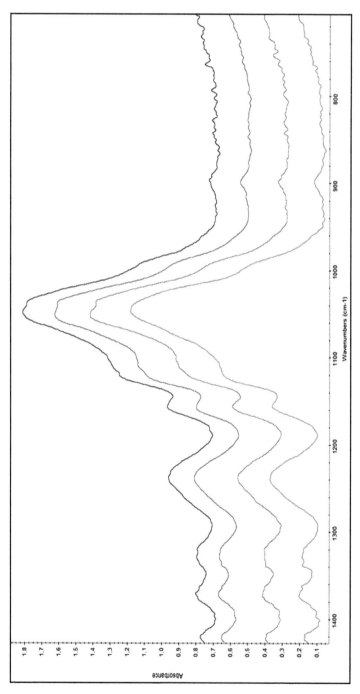

FIGURE 15.5 FTIR spectra of beech wood treated by RFD plasma: the red color—untreated beech wood, the green color—plasma-treated for 20 s, the purple color—plasma-treated for 60 s, and the blue color—plasma-treated for 120 s.

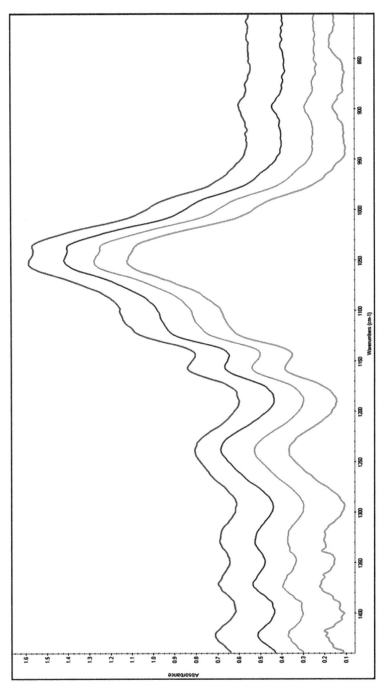

FIGURE 15.6 FTIR spectra of maple wood treated by RFD plasma: the red color—untreated maple, the yellow color—plasma-treated for 20 s, the green color—plasma-treated for 60 s, and the purple color—plasma-treated for 120 s.

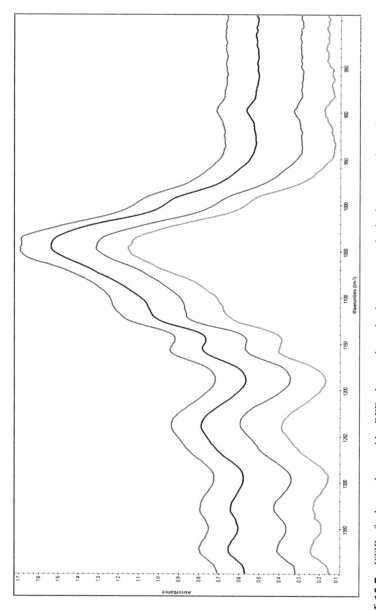

FIGURE 15.7 FTIR of ash wood treated by RFD plasma: the red color—untreated ash, the green color—plasma-treated for 20 s, the blue color—plasma-treated for 60 s, and the purple color—plasma-treated for 120 s.

CHAPTER 9

THE CHEMISTRY OF NONAQUEOUS SOLVENTS: REACTIONS AND CHARACTERISTICS

SONIA KHANNA[*]

Department of Chemistry, School of Basic Science and Research, Sharda University, Greater Noida, India

[*]E-mail: sonia.khanna@sharda.ac.in

ABSTRACT

For last three decades, water has been an ideal solvent for ionic compounds. Reactions susceptible to hydrolytic attack require the nonaqueous solvents. These solvents are gaining importance in view of their suitability for reactions in otherwise impossible reactions. They are selected on the basis of certain parameters such as dielectric constant and solubility. The chemical reactions and properties of various nonaqueous solvents are discussed in this chapter.

9.1 INTRODUCTION

Solvents are an important part of chemical reactions. The chemistry of majority of reactions is dependent on the nature of solvents. Solvents depending on their aqueous or nonaqueous nature determine the fate of reactions. Most of the inorganic chemical reactions familiar to us take place in aqueous solution, but all of them may not favor water as reaction medium. Thus, it becomes important to look out for some other solvent options other than water and number of nonaqueous solvents such as liquid NH_3,[1–2] liquid SO_2,[3–4] methanol,[5] etc. Mixed solvents are also used frequently in acid–base titrations and in studying solute solvent interactions.[6] Mixture of acetic anhydride with nitromethane,[7] acetonitrile,[8] and chloroform[9] are used to study weak bases. Lot of research has been reported on the application of nonaqueous solvents in biotechnology and biocatalysis. Biocatalysis in aqueous media offers unique capabilities and plays a major role in biotransformation technologies.[10–14] The ligand substitution reactions have been investigated using nonaqueous solvents by Nuclear Magnetic Resonance (NMR) methods.[15] The comparison of stabilization, solubilization of magnetic nanoparticles in water and nonaqueous solvents is crucial in nanotechnological applications.[16–17] Current research is focused on use of nonaqueous solvents in electrochemistry for operation of Li ion batteries.[18]

The solubility of a solute in solvent depends on certain factors such as donor–acceptor reaction. Ionic compounds are more soluble in solvents with more donor atoms. Steric factors also play a major role in solubility. Another important factor in solubility of solute in a solvent is the dielectric constant of the solvent. The solvents which have low dielectric constants, possess residual attractive forces of the ions for each other and greater is the tendency of ions to agglomerate into ion pairs. Gutmann reviewed the effect of salvation of lewis acids in inorganic and organic solvents.[19]

In this chapter, the reactions and properties of some of the common nonaqueous solvents liquid NH_3, liquid SO_2, BF_3, H_2SO_4, HF, and N_2O_4 are discussed along with classification of solvents. The solubility of solute in nonaqueous solvents is also discussed.

9.1   CHARACTERISTICS OF SOLVENTS

A solvent is characterized by the following certain properties:

(1) Melting point and boiling point: It indicates the liquid range for any solvent. For example, for water, it is 0–100°C, while for liquid ammonia, it is −77.7°C to −33.5°C.

(2) Heat of vaporization: The degree of association of molecules in a liquid determines its solvent properties. Some information about the nature and strength of associative forces in a molecule can be determined by its heat of vaporization. The higher values indicate higher intermolecular forces. Trouton constant is used to account for intermolecular forces in normal liquids. It is a ratio of heat of vaporization to the boiling point of the solvent. Polar liquids undergo association and have high value of Trouton constant.

Heat of vaporization is defined as heat required for conversion of 1 mole of a liquid at its boiling point completely into vapor. For normal liquids, it is 21.5. A higher value indicates the aggregation of solvent molecules as clusters or aggregates.

(3) Chemical solubility and nature: The behavior of a solute in any solvent also decides its course of action as solute tends to dissolve in a solvent which is chemically similar to it. Polar solute will be dissolved in polar solvent and nonpolar solute in nonpolar solvent.

(4) Dipole moment: Greater is the polarity of solvent or higher the dipole moment of the solvent, greater is the solvation energy released and greater is the solubility of a solute. In general, an ionizing solvent has high values of both dipole moment and dielectric constant.

9.2 CLASSIFICATION OF SOLVENTS

Solvents can be classified in number of ways depending on their physical or chemical properties.

(A) On basis of proton acceptor–proton donor properties.

(i) **Protic or protonic solvents:** They have hydrogen in their composition and are of two types.

(a) acidic or protogenic solvents having a strong tendency to donate protons. For example, CH_3COOH, HCN, etc.

(b) basic or protophilic solvents having a strong tendency to accept protons. For example, NH_3, amines, etc.

(ii) **Aprotic or nonprotonic solvents:** They may or may not have protons in their formula and neither donate or accept protons. For example, C_6H_6, CCl_4, BrF_3, etc.

(iii) Amphiprotic or amphoteric solvents: These solvents can donate or accept protons depending on the nature of reacting species. For example, CH_3COOH, H_2O, NH_3. They undergo autoionization in which proton transfer between two similar neutral molecule take place and a cation–anion pair is formed.

$$H_2O + H_2O \Leftrightarrow H_3O^+ + OH^-$$
$$NH_3(l) + NH_3(l) \Leftrightarrow NH_4^+ + NH_2^-$$
Acid   base   Acid   base
$$NH_3 + 2NH_3 \leftrightarrows NH_2^-$$
$$2NH_3 + 2NH_3 \leftrightarrows N_3^-$$

(B) On the basis of polarity of solvents

(i) Polar, ionizing, or ionic solvents: These solvents have high polarity and can undergo self ionization. They also possess high dielectric constants. They have a high tendency to undergo association leading to an increase in boiling point and hence the liquid range of solvents, for example, H_2O, NH_3, HF, etc.

$$H_2O + H_2O \leftrightarrows H_3O^+ + OH^-$$
$$NH_3 + NH_3 \leftrightarrows NH_4^+ + NH_2^-$$
$$HF + HF \leftrightarrows HF_2 + F^-$$

(ii) Nonpolar or nonionizing solvents: These solvents are nonpolar and do not ionize at all. They have very less associative tendency and do not undergo self-ionization. For example, C_6H_6, CCl_4, etc.

(C) Aqueous and nonaqueous solvents

All the solvents other than water are known as nonaqueous solvents. For example, liquid NH_3, glacial CH_3COOH, HF, liquid $CHCl_3$, unhydrous H_2SO_4, dinitrogen tetraoxide, bromine trifluoride, DMSO. Nonaqueous solvents are often selected on the basis of their liquid range and relative permittivity. The behavior of various nonaqueous solvents is also discussed. The physical properties of some of the common nonaqueous solvents are summarized in Table 9.1.

TABLE 9.1 Physical Properties of Some Common Nonaqueous Solvents.

Solvent	m.p.°C	b.p. °C	Dipole moment D	Dielectric constant
H_2O	0.0	100	1.85	78.5
NH_3	−77.7	−33.4	1.47	22.4

TABLE 9.1   *(Continued)*

Solvent	m.p.°C	b.p. °C	Dipole moment D	Dielectric constant
HF	−83	19.4	1.9	83.6
N_2H_4	2.0	113.5	1.83	51.7
N_2O_4	−11.2	21.5	-	2.42
CH_3OH	−97.8	65.0	1.08	31.6
H_2SO_4	10.4	338	-	100
HCN	−13.4	25.7	2.8	114.9
$(CH_3CO)_2O$	−71.3	136.4	2.8	20.5

Some of the common nonaqueous solvents are discussed below.

(A) Liquid NH_3: It is the most extensively studied used solvents in organic and inorganic reactions. It differs from water in terms of its dielectric constant. The dielectric constant of liquid NH_3 (22) is less than that of water (78.5). The lower dielectric constant results in decreased ability to dissolve ionic compounds. However, lower viscosity of liquid NH_3 is expected to promote greater ionic mobilities and overcome the dielectric constant. The properties of ammonia are given in Table 9.2.

TABLE 9.2   Physical Properties of Liquid Ammonia.

Boiling point	−33.38°C
Freezing point	−77.7°C
Density	0.725 g cm^3
Dielectric constant	22.0 at −33.5°C
Specific conductance	1×10^{-11} Ω^{-1} cm^{-1}
Viscosity	0.254 g cm^{-1} s^{-1}
Ion product constant	5.1×10^{-27} mol^2 L^{-2}

The ammonia molecule possesses some degree of polarity, resulting from pyramidal structure of the molecule. It is sp^3 hybridized and has the total dipole moment of 1.46D. It is used as solvent for nonpolar molecules. It shows reasonable liquid range (−77°C to −33°C), easily liquefied with dry ice and can be stored in thermos flask. The melting point and boiling point of ammonia are abnormal with respect to PH_3, AsH_3, and SbH_3, due to intermolecular association of hydrogen bonding. It is good solvent for organic molecule like amines, benzene, etc. than water but a poor solvent for inorganic molecules. However, this difference can be utilized for separation of unstable compounds. Ammonia is highly hygroscopic and liquid ammonia system must be protected from atmospheric moisture, and are therefore reactions are carried out in sealed tubes. Like water, liquid ammonia is also associated through hydrogen bonding. But the N-H...N bond is weaker than OH....O bond. The properties based on association are less pronounced in liquid ammonia than in water.

9.3   FEATURES OF LIQUID NH$_3$

(i)   It is less viscous than water.
(ii) It has high critical temperature and pressure.
(iii) It is less associated than water (due to lesser hydrogen bonding).
(iv) It forms metal–ammonia solutions with alkali metals.
(v)   It is a poor conductor of electricity.
(vi) Specific heat of ammonia is greater than water.

9.4   CHEMICAL REACTIONS IN LIQUID NH$_3$

1. **Precipitation reactions:** The reactions in which precipitation occurs on mixing two solutions are called metathetical or precipitation reactions. Due to differences in physical properties of liquid solvents, various substances differ in their solubilities in these solvents. For example, in H$\square$O, silver nitrate and barium chloride react to give a precipitate of silver chloride leaving barium nitrate in the solution.

$$2AgNO\square + BaCl\square \rightarrow 2AgCl \downarrow + Ba(NO\square)\square$$

Many precipitation reaction that are normally not possible in water may take place in liquid NH$_3$, for example,

$$KCl + AgNO_3 \rightarrow AgCl + KNO_3$$

$$KNO_3 + AgCl \rightarrow KCl + AgNO_3$$

$$KI + NH_4Cl \rightarrow KCl + NH_4I$$

Precipitation reactions involve double decomposition because of the differences in solubilities.

(i) Only a few chlorides ($NaCl$, NH_4Cl, $BeCl_2$) are soluble in liquid ammonia, while other halides are insoluble. Hence, many chlorides can be precipitated by mixing any soluble metal salt with ammonium chloride in liquid ammonia

$$KI + NH_4Cl \rightarrow KCl \downarrow + NH_4I$$

$$LiNO_3 + NH_4Cl \rightarrow LiCl \downarrow + NH_4NO_3$$

White ppt of BaCl☐ is produced when silver chloride and liquid ammonia brought together.

$$2AgCl + Ba(NO☐)☐ \leftrightarrow BaCl☐ \downarrow + 2AgNO☐$$

2. Acid–base reactions: The neutralization reactions in liquid NH_3 differs from that in aqueous medium. In aqueous medium, neutralization reaction results information of salt and water, while in liq NH_3, salt and ammonia are formed, for example,

$$NH_4Br + KNH_2 \rightarrow KBr + 2NH_3$$

In liquid NH_3, the substance which increases the concentration of NH_4^+ in liquid ammonia are known as ammono acid while the substance which increase the concentration of NH_2^-, NH^{2-}, or N^{3-} ions in liquid ammonia are called as ammono base.

Ammono acids undergo different types of reactions
i) Neutralization reactions: $KNH_2 + NH_4Cl \rightarrow KCl + 2NH_3$
ii) With active metals: Ammonium salts in liquid ammonia can react with active metals.

$$Co + 2NH_4NO_3 \rightarrow Co(NO_3)_2 + 2NH_3 + H_2$$

iii) Protolysis: Compounds incapable of donating protons in water, readily donate protons in liquid ammonia. For example, urea, acetamide and sulphamide.

$$NH_2CONH_2 + NH_3 \rightarrow {}^-HNCONH_2 + NH_4^+$$

$$CH_3CONH_2 + NH_3 \rightarrow CH_3CONH^- + NH_4^+$$

$$H_2NSO_2NH_2 + NH_3 \rightarrow H_2NSO_2NH^- + NH_4^+$$

Ammono bases are used to precipitate amides, imides and nitrides of many metals.

3. Solvolysis reactions: The mechanism of solvolysis resembling hydration in water results in the attachment of solvent molecules to the cation, the anion, or the molecule of the solute as whole through coordination or ion–dipole interactions. Solvolysis in liquid ammonia is known as ammonolysis. Amide, imide, and nitride are usual products of ammonolysis. For example,

i) Organic Halides undergo slow ammonolysis to form amines:

$$RX + 2NH_3 \rightarrow RNH_2 + NH_4X$$

$$2RX + 3NH_3 \rightarrow R_2NH + 2NH_4^+ + 2X^-$$

$$3RX + 4NH_3 \rightarrow R_3N + 3NH_4^+ + 3X^-$$

For example,

$$AlCl_3 + 2NH_3 \rightarrow AlCl_2(NH_2) + NH_4^+ + Cl^-$$

$$BX_3 + 6NH_3 \rightarrow B(NH_2)_3 + 3NH_4^+ + 3X^-$$

ii) Hydrides of alkali and alkaline earth metals are ammonialyzed to corresponding metal amide and hydrogen:

$$MgH_2 + 2NH_3 \rightarrow Mg(NH_2)_2 + 2H_2$$

$$NaH + NH_3 \rightarrow NaNH_2 + H_2$$

iii) Oxides of alkali metals are ammonialyzed to metal amide and hydroxides.

$$K_2O + NH_3 \rightarrow KOH + KNH_2$$

4. Solutions in liquid ammonia: The most striking property of liquid ammonia is its ability to dissolve alkali metals. If a small piece of alkali metal is put in liquid ammonia, the resulting solution acquires a deep blue color. On addition of more alkali metal in solution, a bronze colored phase separates out and floats on the surface. This solution exhibits good electrolytic conductivity which can be explained by the dissociation into cations and anions. When alkali metals are dissolved in liquid ammonia, they ionize to give metal ions and valence electrons as:

$$Na \leftrightarrow Na^+ + e^-$$

Both alkali metal and electron become solvated by ammonia molecules.

$$Na^+ + xNH_3 \rightarrow [Na(NH_3)x]^+ \text{ (ammoniated cation)}$$

The solution contains large amount of ammoniated electrons which accounts for its conductivity.

$$e^- + yNH_3 \rightarrow [e(NH_3)y]^- \text{ (amnoniated electron)}$$

The ammoniated electrons are responsible for blue color of solution. In liquid ammonia, many compounds react with excess of ammino ions to form soluble complexes. For example,

$$Zn(OH)_2 + 2NaNH_2 \rightarrow Na_2[Zn(NH_2)_4]$$

$$Zn(NO_3)_2 + 4KNH_2 \rightarrow K_2[Zn(NH_2)] + 2KNO_3 + 2NH_3$$

$$AgNH_2 + KNH_2 \rightarrow K[Ag(NH_2)_2]$$

$$AlCl_3 + 4NaNH_2 \rightarrow Na[Al(NH_2)]_4 + 3NaCl$$

9.5 SOLUBILITIES IN LIQUID AMMONIA

(i) Organic compounds: Organic compounds like alcohol, halogen compounds like chloroform, esters, ketones, phenol and its derivatives are soluble in liquid NH_3.

(ii) Inorganic compounds: Liquid NH_3 is a poor solvent for inorganic compounds like H_2O, due to low dielectric constant of NH_3 as compared with H_2O. Electron rich compounds like iodine compounds and nonpolar substance like hydrocarbons are however soluble in liquid NH_3. Nitrates, thiocynates, perchlorates, and most cyanides are also soluble along with oxides, hydroxides, sulphates, carbonates, phosphates, sulphites, and sulphides. Solubility of halides follows the trend: $I^- > Br^- > F^-$ Only NH4+, Be^{2+} and Cl⁻ are soluble in liquid NH_3.

(iii) Nonmetals like P, S, and I dissolve in liquid NH_3 and react with it.

9.8 LIQUID SO$_2$

It is a nonprotonic solvent or aprotic solvent because it does not contain any hydrogen atom. At room temperature, it is a gas, but can be readily liquefied and kept in liquid state.

It is also one of the important nonaqueous solvent and widely used in industry. It is a gas under normal temperature and pressure and can be liquefied easily. Its liquid range is quite high (−10–75.5°C). Unlike liquid ammonia, it can be used as solvent for many covalent substances. It shows low dielectric constant (17.4 at −20°C) and is a poor solvent for ionic compounds but highly charged species are soluble in it due to generation of vander Waal interactions by π electrons in SO_2. Resulting solutions are highly conducting in nature. The physical properties are given in Table 9.3.

TABLE 9.3 Physical Properties of Liquid SO$_2$.

Boiling point	−10°C
Freezing point	−75.5°C
Density	0.725 g cm³
Dielectric constant	17.4.0 at −20°C
Specific conductance	$3.4 \times 10^{-8}\ \Omega^{-1}\ cm^{-1}$
Viscosity	$0.428\ g\ cm^{-1}\ s^{-1}$

9.9   CHEMICAL REACTIONS IN LIQUID SULFUR DIOXIDE

1.   Precipitation Reactions: Like NH_3, several insoluble materials can be precipitated in SO_2 due to selective solubility of compounds. For example, metal halides are precipitated from solution in liquid SO_2.

$$AlCl_3 + 3NaI \rightarrow 3NaCl\downarrow + AlI_3$$

$$SbCl_3 + 3LiI \rightarrow SbI_3\downarrow + LiCl$$

$$BaI2 + Zn(CNS)_2 \rightarrow Ba(CNS)_2\downarrow + ZnI_2$$

2.   Acid–base reactions: Like NH_3, SO_2 also self ionize and forms SO^{2+} and SO_3^{2-} ions.

$$SO_2 + SO_2 \leftrightarrows SO^{2+} + SO_3^{2-}$$

The compounds which increase the concentration of SO^{2+} act as acids and those which increase the concentration of SO_3^{2-} act as base in liquid SO_2.
For example,

$$SOCL_2 \leftrightarrows SO^{2+} = 2Cl^- \quad \text{(Acidic)}$$

$$CsSO_3 \leftrightarrows 2Cs + SO_3^{2-} \text{ (Basic)}$$

Neutralization reaction: $SOCl_2$ (Acid) + $CsSO_3$ (Base) $\rightarrow$ $2CsCl$ + $2SO_3$

3.   Solvolysis reactions: The solvolytic reactions in liquid SO_2 are not as common as they are in other protic solvents such as NH_3, H_2O. For example,

$$WCl_6 + SO_2(l) \rightarrow WOCl_4 + SOCl_2$$

$$UCl_6 + 2SO_2(l) \rightarrow UO_2Cl_2 + 2SOCl_2$$

$$PCl_5 + SO_2(l) \rightarrow POCl_2 + SOCl_2$$

$$2CH_3COONH_4 + 2SO_2 \rightarrow (NH_4)_2SO_3 + (CH_3COO)_2SO$$

$$(CH_3COO)_2SO \leftrightarrows SO_2 + (CH_3CO)_2O$$

Liquid SO_2 behaves as Lewis base due to presence of lone pairs on S and O and forms a variety of solvates like $AlCl_3.2SO_2$, $K(SCN).2SO_2$, dioxane.$2SO_2$, etc.

4. Complex formation: SO_2 forms the complex due to availability of two pairs of electrons on both S and O. Many metal salts can react with SO_3^{2-} to form precipitates which are soluble in excess of SO_3^{2-}. For example,

$$2AlCl_3 + 3K_2SO_3 \rightarrow Al_2(SO_3)_3 + 6KCl$$

$$Al_2(SO_3)_3 + 3K_2SO_3 \rightarrow 2K_3[Al(SO_3)_3]$$

Similar behavior has been shown by other salts also.

5. Organic reactions: A large number of organic compounds are soluble in liquid SO_2 because of its inert nature. Some of the reactions are:

Sulphonation: $C_6H_6 + SO_2 \,(l) \rightarrow C_6H_5SO_3H$

Friedel Crafts reaction: $C_6H_6 + C_2H_5COCl \xrightarrow[\text{Li}_2, \text{ SO}_2]{\text{AlCl}_3} C_6H_5COC_6H_5 + HCl$

Solubility in Liquid SO_2: Iodides of alkali metals are soluble in liquid SO_2, while most of ammonium and mercuric salts are highly insoluble. Aliphatic hydrocarbons are insoluble while aromatic hydrocarbons and alkenes are highly soluble. This makes liquid SO_2 an ideal solvent for organic reactions. Sulfur dioxide possessing low dielectric constant suggests that this solvent possess limited potentialities toward dissolution of various solutes. On the other hand, it possesses some unusual capabilities toward covalent halides existing as electrolyte in liquid sulfur dioxide. On account of low dielectric constant, great bulk of solute is present in the form of ion pairs.

9.10   SPECIAL FEATURES OF LIQUID SO_2

i)   It has low boiling point allowing its easy evaporation and thus removal from the reaction mixture.

ii) Liquid SO_2 is an inert solvent.

iii) It shows unusual solvating powers for both covalent and electrovalent compounds.

iv) It is used as diluents for superacid medium ($CHSO_3F–SbF_5$) which helps in reducing its viscosity without affecting its acidity.

9.11   LIQUID HYDROGEN FLUORIDE

It resembles water in terms of polarity and solubility of ionic solids. Because of the presence of electronegative F^-, liquid HF is an excellent ionizing solvent. It has a liquid range comparable to water and shows extensive hydrogen bonding. Due to its high reactivity, it cannot be stored in glass containers so it is stored in inert containers made up of Teflon. It persists in gas phase in the form of chains and rings. Unlike other nonaqueous solvents, liquid HF have high dielectric constant (83.6 at 0°C) and dipole moment of 1.90D. It possesses wide liquid range as compared with other nonaqueous solvents.

Liquid HF has equivalent conductance of 1.4×10^{-5} ohm^{-1} lower as compared with water (6.0×10^{-8} ohm^{-1}). Thus, both dipole moment and dielectric constant are comparable to water indicating it suitability as a solvent for inorganic reactions. It is quite poisonous and dissolves only few substances, limiting its use as nonaqueous solvent. The physical properties of liquid HF are given in Table 9.4.

TABLE 9.4   Physical Properties of Liquid HF.

Boiling point	+19.4°C
Freezing point	−89.4°C
Density	0.99 gcm^3
Dielectric constant	83.6 at −20°C
Specific conductance	1.4×10^{-5} Ω^{-1} cm^{-1}
Viscosity	0.256 g cm^{-1} s^{-1}
Dipole moment	1.90D

9.11.1   CHEMICAL REACTIONS IN LIQUID HF

i) Acid–base reactions self-ionization: Like NH_3 and SO_2, HF also undergoes self ionization as
$$HF + HF \rightarrow H_2F^+ + F^-$$
Fluoronium ion

Thus substance which increase the concentration of H_2F^+ will act as acid and those which increase the concentration of F^- will act as base.

Electron acceptor species such as BF_3, SnF_4, SbF_5 act as acid in HF

$$BF_3 + 2HF \rightarrow H_2F^+ + BF_4^-$$

$$SnF_4 + 4HF \rightarrow SnF_6^{2-} + 2H_2F^+$$

$$SbF_5 + 2HF \rightarrow H_2F^+ + SbF_6^-$$

Acids behave as base in HF. For example,

$$HNO_3 + HF \rightarrow H_2NO_3^+ + F^-$$

$$H_2SO_4 + HF \rightarrow H_2SO_4^+ + F^-$$

In addition, liquid HF converts some of the oxides to oxyhalides. For example,

$$5HF + MnO_{4^-} \rightarrow MnO_3F + H_3O^+ + 2HF^{2-}$$

ii) Precipitation reactions: Compounds like periodates, sulphates, and perchlorates are easily precipitated in liquid HF. For example,

$$AgF + KIO_4 \rightarrow AgIO_4 + NaF$$

$$2AgF + Na_2SO_4 \rightarrow Ag_2SO_4 + 2NaF$$

$$TlF + KClO_4 \rightarrow TlClO_4 + KF$$

iii) Addition compounds: Addition compounds are formed between metallic fluorides, for example, KF.2HF, KF.3HF, etc.

iv) Solvolysis reactions: Salts and acids undergo solvolysis in HF to form F-, for example,

$$KCN + HF \rightarrow HCN + K^+ + F^-$$

$$KCl + HF \rightarrow HCl + K^+ + F^-$$

$$KNO_3 + HF \rightarrow HNO_3 + K^+ + F^-$$

$$K_2SO_4 + 2HF \rightarrow H_2SO_4 + 2K = + F^-$$

v) Redox reactions: Metals and hydrofluoro acids undergo reactions in HF similar to reactions in aqueous medium.

9.12 LIQUID DINITROGEN TETRAOXIDE N_2O_4

Liquid dinitrogen tetraoxide N_2O_4, is a protic nonaqueous solvent. It can be easily prepared and has convenient liquid range (-11.2–$21.1°C$). Like other nonaqueous solvents, it has a very low dielectric constant (2.42) and is a poor solvent for polar substance and a good solvent for nonpolar substance. However, it shows very poor self-ionization and low specific self conductance (1.3×10–13 ohm^{-1}cm^{-1} at $10°C$). The physical characteristics are given in Table 9.5.

TABLE 9.5 Physical Properties of N_2O_4.

Boiling point	$294.2°C$
Melting point	$261.8°C$
Density	1.49 g cm^3 (at 273 K)
Relative permittivity	2.42 (at 291 K)

9.12.1 CHEMICAL REACTIONS IN LIQUID N_2O_4

1. Acid–Base reactions: The self-ionization of N_2O_4 can be given as:

$$2N_2O_4 \leftrightharpoons 2NO^+ + 2NO_3^-$$

Any substance that furnishes the NO^+ ion behave as acid and those furnishing NO_3^- would behave as base in liquid N_2O_4. Hence, NOCl, NOBr, etc. behave as acid in N_2O_4, while $AgNO_3$ behave as base in N_2O_4.

$$NOCl + AgNO_3 \rightarrow AgCl + N_2O_4$$

2. Solvolytic reactions: Many compounds have been solvated in N_2O_4 to form nitrates. For example, Li_2CO_3 in N_2O_4 is solvolysed to form lithium nitrate and N_2O_3.

$$Li_2CO_3 + 2N_2O_4(l) \rightarrow 2LiNO_3 + N_2O_3 + CO_2\uparrow$$

$$KCl + N_2O_4(l) \rightarrow NOCl + KNO_3\downarrow$$

Thus, liquid N_2O_4 are used to prepare anhydrous nitrates of metals which are otherwise difficult to obtain. For example, in case of metal carbonyls.

$$Mn_2(CO)_{10} + N_2O_4 (l) \rightarrow Mn(CO)_5NO_3 + Mn(CO)_4NO + CO.$$

9.13   BROMINE TRIFLUORIDE BrF_3:

It is aprotic nonaqueous solvent. It is a pale yellow colored liquid at 298K. It is an extremely strong fluorinating agent and can fluorinate anything that dissolves in it. It is stored in quartz container. The physical properties are given in Table 9.6.

TABLE 9.6   Physical Properties of BrF_3.

Boiling point	408°C
Melting point	281.8°C
Density	2.49 g cm³
Relative permittivity	107
Self-ionization constant	8.0×10^{-3} (at 281.8 K)

Reactions in BF_3

i) **Acid–Base Reactions:** BF_3 acts as strong lewis acid and readily accepts F- alkali metal fluorides, BaF_2 and AgF in BrF_3 gives salts containing BrF^{4-} ion, that is, $Ba[BrF_4]_2$ and $Ag[BrF_4]$. On reaction with more stronger F⁻ acceptor, salts containing $[BrF_2]^+$ are formed. For example,

$$SbF_5 + BF_3 \rightarrow [BrF_2]^+ + [SbF_6]^-$$

$$SnF_4 + BF_3 \rightarrow [BrF_2]^+ + [SnF_6]^-$$

ii) **Other reactions in BF_3:** Many reactions requiring fluorination can be carried out in BrF_3. For example,

$$Ag + Sb \rightarrow Ag[SbF_6]$$
$$(\text{liquid } BF_3)$$
$$2KCl + Sn \rightarrow K_2[SnF_6]$$
$$(\text{liquid } BF_3)$$

9.14 ANHYDROUS SULFURIC ACID (H_2SO_4)

Anhydrous sulfuric acid is widely used non aqueous solvent. It is liquid at 298K and has a long liquid range. It has high boiling point and high viscosity. It is used as dehydrating agent to extract water from chemical compounds. The physical parameters are given in Table 9.7.

TABLE 9.7 Physical Properties of H_2SO_4.

Boiling point	$\approx 603°C$
Melting point	$283.4°C$
Density	1.84 g cm^3
Relative permittivity	110 (at 292 K)
Self ionization constant	2.7×10^{-4} (at 298 K)

9.14.2 CHEMICAL REACTIONS IN H_2SO_4:

i) Acid–base behavior: Sulfuric acid undergoes self-ionization as:

$$H_2SO_4 + H_2SO_4 \rightarrow H_3SO_4^+ + HSO_4^-$$

Substance increasing the concentration of $H_3SO_4^+$ are regarded as acid and the substance which increase the concentration of HSO_4^- are regarded as base in H_2SO_4. These ions carry electric current by proton switching mechanism.

For example, $HClO_4 + H_2SO_4 \rightarrow ClO_4^- + H_3SO_4^+$
$$SO_3 + 2H_2SO_4 \rightarrow H_3SO_4^+ + H_2S_2O_7^-$$
$$HNO_3 + 2H_2SO_4 \rightarrow NO_2^+ + H_3O^+ + 2H_2SO_4^-$$
$$H_3BO_3 + 6H_2SO_4 \rightarrow B(HSO_4)_4^- + 2HSO_4^- + 3H_3O^+$$

Sulfuric acid is a strong acid and reacts with almost all chemical species forming sulphates and hydrogen sulphates. Strong acids like perchloric acid behave as an electrolyte in H_2SO_4. Only a few substance like disulfuric acid behave as acid in sulfuric acid.

$$SO_3 + H_2SO_4 \rightarrow H_2S_2O_7$$
$$H_2S2O_7 + H_2SO_4 \rightarrow H_3SO_4^+ + HS_2O_7^-$$

Substances with lone pairs of electrons such as amides behave as base in H_2SO_4 and those which dissociate to produce HSO_4^- behave as base in H_2SO_4.

$$H_2NCONH_2 + H_2SO_4 \rightarrow H_2NCONH_3^+ + HSO_4^-$$
$$RCOOH + 2H_2SO_4 \rightarrow CH_3COOH^{2+} + HSO_4^-$$

ii) Redox reactions: Anhydrous sulfuric acid is a mild oxidizing agent. Colored solutions are obtained on treatment of chalcogens with sulfuric acid. For example, S_8^{2+} (blue), S_{16}^{2+} (red) and Se_8^{2+} (green).

iii) Solvolysis reactions: Many organic compounds like nitriles, nitro compounds, sulphones and sulphoxides undergo slow solvolysis in H_2SO_4.

$$RCN + H_2SO_4 \rightarrow RCNH^+ + HSO_4^-$$
$$RCNH^+ + HSO_4^- \rightarrow RCONH_2 + SO_3$$

9.15   CONCLUSION

Nonaqueous solvents have emerged as ray of light in case of impossible reactions. They are widely used as reaction medium in various organic and inorganic reactions. Many nonaqueous solvents have shown better performance as solvent than conventional solvents. The studies can extend to other interdisciplinary areas such as medical biotechnology, etc.

KEYWORDS

- dielectric constant
- solubility in nonaqueous solvents
- solvolysis
- acid–base reactions

REFERENCES

1. Franklin, E. C. Reactions in Liquid Ammonia. *J. Amer. Chem. Soc.* **1905,** *27*, 820.
2. Jander, G. Chemistry in Non-aqueous Solvents. In *Vieweg and Interscience*; Addison, C. C.; Beerunschweig: New York, 1966.
3. Waddington, T. C. *Nonaqueous Solvent System.* Academic Press: London, New York, 1966.
4. Elving, P. J.; Markowitz, J. M. Voltametry in Liquid Sulphur Dioxide–Techniques and Theoretical Problems. *J. Phys. Chem.* **1961,** *65*, 680–686.
5. Luginnia, J.; Suta, K.; Turks, M. Application of Liquid Sulfur Dioxide as a Solvent for Organic Transformations. *20th European Symposium on Organic Chemistry*, Germany, Cologne, 2–6 July, 2017
6. Fuoss, R. N. *J. Chem. Edu.* **1955,** *32,* 527.
7. Pearson, R. G.; Vogelsong. Acid Base Equilibrium Constants for 2,4- Dinitrophenol and Some Aliphatic Amines in Non Aqueous Solvents. *J. Am. Chem. Soc.* **1958,** *80* (5), 1038–1043.
8. Marcus, R. A.; Winkler, C. A. Studies on RDX and Related Compounds: Analysis for Nitric Acid in Acetic Acid–Acetic Anhydride Media. *Can J. Chem.* **1953,** *31*, 214–215.
9. Salvesin, B. Med. Vorsk. farm, Slsk, 20, 21 (1958).
10. Khmelnitsky; Rich, J. D. Biocatalysis in Nonaqueous Solvents. *Curr. Opin. Chem. Biol.* **1999,** *3* (1), 47–53.
11. Jemil, S.; Ayadi-Zouari, D.; Hlima, H. B.; Bejar, S. Biocatalyst: Applications and Engineering for Industrial Purposes. *Crit. Rev. Biotechnol.* **2014,** *6* (2), 1–13.
12. Watt, G. Non Aqueous Solvents: Media for Chemical Reactions. *J. Chem. Educ.* **1954,** *31* (4), 220.
13. Chen, K.; Arnold, F. H. Enzyme Engineering for Nonaqueous Solvents: Random Mutagenesis to Enhance Activity of Subtilisin E in Polar Organic Media. *Biotechnology* **1991,** *9*, 1073–1077.
14. Gupta, A.; Khare, S. K. Enzymes from Solvent-tolerant Microbes: Useful Biocatalysts for Non-aqueous Enzymology. *Crit. Rev. Biotechnol.* **2009,** *29* (1), 44–54.
15. Moore, P. Probing Ligand Substitution Reactions in Nonaqueous Solvents by NMR Methods: Stopped Flow and High Pressure Flow Fourier Transform NMR and Neutral Abundance 17o Line Broadening. *Pure Appl. Chem.* **1985,** *57* (2), 347–354.

16. Kharisova, O. V.; Kharisov, B. L.; Ortiz, E. G. Dispersion of Carbon Nanotubes in Water and Nonaqueous Solvents. *RSC Advances* **2013,** *3*, 24812–24852.

17. Porter, G. B.; Hoggard, P. E. Photochemistry of Ru(II) Complexes in Nonaqueous Solvents. *J. Photo Chem.* **1985,** *5*, 165–165.

18. McCloskey, B. D. M.; Bethune, D. S.; Shelby, R. M.; Kumar, G. G.; Luntz, A. C. Solvents Critical Role in Non Aqueous Lithium–Oxygen Battery Electrochemistry. *J. Phys. Chem.* **2011,** *2* (10), 1161–1166.

19. Gutmann, V. *New Pathways in Inorganic Chemistry.* Cambridge University Press, 1968; pp 65–87.

CHAPTER 10

FEATURES OF SELECTIVE SORPTION OF LANTHANUM FROM SOLUTION CONTAINING IONS OF LANTHANUM AND CERIUM BY INTERGEL SYSTEM HYDROGEL OF POLYMETHACRYLIC ACID— HYDROGEL OF POLY-2-METHYL-5-VINYLPYRIDINE

T. K. JUMADILOV* and R. G. KONDAUROV

JSC "Institute of Chemical Sciences after A.B. Bekturov," Almaty, Republic of Kazakhstan

Corresponding author. E-mail: jumadilov@mail.ru

ABSTRACT

The study is a short review devoted to features of selective sorption of lanthanum ions from the solution, which contains lanthanum and cerium ions. It is found that specific electric conductivity increases and concentration of hydrogen ions decreases, during mutual activation of polymethacrylic acid (PMAA) and poly-2-methyl-5-vinylpyridine (P2M5VP) hydrogels in intergel system on their basis in an aqueous medium. Also, it should be noted that a significant increase in the swelling degree of PMAA and P2M5VP is observed in water. Sorption of lanthanum ions by intergel system polymethacrylic acid hydrogel (hPMAA)–poly-2-methyl-5-vinylpyridine hydrogel (hP2M5VP) is accompanied by radically different changes of electrochemical and conformational properties of polymer hydrogels. There is a decrease of specific electric conductivity, pH, and swelling degree due to the folding of polymer globe in process of lanthanum ions sorption. Mutual activation with a further transition into a highly ionized state provides much higher values of lanthanum ions extraction degree of the intergel system comparatively with initial hydrogels of PMAA and P2M5VP. Maximum sorption of lanthanum is observed at 50% hPMAA–50% hP2M5VP ratio, sorption degree is 89.65%. Polymer chain binding degree (in relation to lanthanum ions) has maximum values at hydrogels ratio 50% hPMAA–50% hP2M5VP, the value of the binding degree is 74.67%. Intergel system hPMAA–hP2M5VP have maximum values of cerium ions extraction degree at 67% hPMAA–33% hP2M5VP (sorption degree is 87.67%). Polymer chain binding degree at this ratio also has the highest values; it is 72.72%. Selective sorption of lanthanum ions by the intergel system 50% hPMAA–50% hP2M5VP indicates that selectivity of the intergel system is manifested to lanthanum ions. Extraction degree of lanthanum ions is 65.40% of cerium ions 24.38%. Polymer chain binding degree is in relation to lanthanum ions—54.25%, in relation to cerium ions—20.22%.

10.1 INTRODUCTION

The main feature of the intergel system is the absence of direct contact between polymer hydrogels in solution.[1–3] In other words, the interaction of polymer hydrogels of acid and basic nature occurs remotely. The intergel system is presented in Figure 10.1. Polyacid and polybases are put in a special glass filter, pores of which is permeable for low-molecular ions but nonpermeable for hydrogels dispersion.

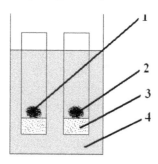

FIGURE 10.1   Intergel system; 1—polyacid, 2—polybases, 3—glass filter, and 4—solution.

During remote interaction of hydrogels, the following chemical reactions occur:

1. Dissociation of –COOH groups of internode links:

$$-COOH \rightarrow COO^- \cdots H^+ \rightarrow -COO^- + H^+$$

It should be mentioned that first there is an occurrence of ionization with ionic pairs formation; after that, ionic pairs partially dissociate on separate ions.

2. The nitrogen atom in pyridine ring is ionized and partially dissociates:

$$\equiv N + H_2O \rightarrow \equiv NH^+ \cdots OH^- \rightarrow \equiv NH^+ + OH^-$$

3. After that, nitrogen atom also interacts with the proton, which is cleaved from the carboxyl group:

$$\equiv N + H^+ \rightarrow \equiv NH^+$$

4. H^+ and OH^- ions, formed in result of the interaction of functional groups with water molecules, form water molecules (it is valid for equimolar concentrations of protons and hydroxyl ions)

$$H^+ + OH^- \rightarrow H_2O$$

As seen from mentioned above processes, in the result of the remote interaction effect of polymer hydrogels, their mutual activation occurs. Mutual activation proposes the transfer of hydrogels into a highly ionized state. The result of this phenomenon is the significant change of electrochemical properties (specific electric conductivity, pH) of solutions and changes of conformational and sorption properties of macromolecules.

In accordance with chemical reaction 1, the dissociation of carboxyl groups on carboxylate anions and protons occurs, and it depends on the degree of dissociation. Due to the association of protons by heteroatoms, the total amount of hydrogen ions in solution decreases, what, in turn, provides additional dissociation of other (nonreacted) functional carboxyl groups. It occurs according to Le Chatelier principle—due to the shift of the equilibrium to right (to the side of protons formation).[4]

These interactions provide the formation of same-charged groups on internode links of both (acid and basic) hydrogels.[5] Due to laws of electrostatics, these groups repulse from each other and provide unfolding of the macromolecular globe. The end result of such electrostatic interactions is a significant increase in the swelling of polymeric macromolecules.[6,7]

10.2   EXPERIMENTAL PART

10.2.1   EQUIPMENT

For measurement of solutions, specific electric conductivity conductometer "MARK-603" (Russia) was used; hydrogen ions concentration was measured on Metrohm 827 pH meter (Switzerland). Samples' weight was measured on analytic electronic scales Shimadzu AY220 (Japan). La^{3+} ions concentration in solutions was determined on spectrophotometers SF-46 (Russia) and Jenway-6305 (The United Kingdom).

10.2.2   MATERIALS

Studies were carried out in 0.005 M 6-water lanthanum nitrate solution. Polymethacrylic acid hydrogel (hPMAA) was synthesized in the presence of crosslinking agent N,N-methylene-bis-acrylamide and redox system $K_2S_2O_8$–$Na_2S_2O_3$ in the water medium. Synthesized hydrogels were crushed into small dispersions and washed with distilled water until the constant conductivity value of aqueous solutions was reached. Poly-2-methyl-5-vinylpyridine (hP2M5VP) hydrogel of Sigma-Aldrich Company (linear polymer crosslinked by divinylbenzene) was used as polybases.

   For investigation task from synthesized hydrogels, an intergel pair hPMAA–hP2M5VP was created. Swelling degrees of hydrogels were: $\alpha_{(hPMAA)} = 20.65$ g/g and $\alpha_{(hP2M5VP)} = 3.20$ g/g.

10.2.3   ELECTROCHEMICAL INVESTIGATIONS

Experiments were carried out at room temperature. Studies of intergel system were made in the following order: each hydrogel was put in separate glass filters, pores of which are permeable for low molecular ions and molecules, but impermeable for hydrogels dispersion. After that, filters with hydrogels were put in glasses with lanthanum nitrate solution. Electric conductivity and pH of overgel liquid was determined in the presence of hydrogels in solutions.

10.2.4 DETERMINATION OF HYDROGELS SWELLING

The swelling degree was calculated according to the equation:

$$K_{sw} = \frac{m_2 - m_1}{m_1}$$

where m_1 is the weight of dry hydrogel and m_2 is the weight of the swollen hydrogel.

10.2.5 METHODOLOGY OF LANTHANUM IONS DETERMINATION

Methodology of lanthanum ions determination in solution is based on the formation of colored complex compound of organic analytic reagent Arsenazo III with lanthanum ions.[8]

Extraction (sorption) degree was calculated by the following equation:

$$\eta = \frac{C_{initial} - C_{residual}}{C_{initial}} \times 100\%$$

where $C_{initial}$ is the initial concentration of lanthanum in solution (g/L) and $C_{residue}$ is the residual concentration of lanthanum in solution (g/L).

Polymer chain binding degree was determined by calculations in accordance with the equation:

$$\theta = \frac{v_{sorb}}{v} \times 100\%$$

where v_{sorb} is the quantity of polymer links with sorbed lanthanum (mol); v is the total quantity of polymer links (if there are two hydrogels in solution, it is calculated as the sum of each polymer hydrogel links) (mol).

Effective dynamic exchange capacity was calculated by the formula:

$$Q = \frac{v_{sorbed}}{m_{sorbent}}$$

where v_{sorbed} is the amount of sorbed metal (mol) and $m_{sorbent}$ is the mass of the sorbent (if there are two hydrogels in solution, it is calculated as the sum of mass of each of them) (g).

10.2.6 MUTUAL ACTIVATION OF HYDROGELS IN INTERGEL SYSTEM hPMAA–hP2M5VP IN AN AQUEOUS MEDIUM

Dependence of specific electric conductivity from molar ratios in time is shown in Figure 10.2. Conductivity increases with time for all ratios of hPMAA:hP2M5VP. The absence of the second hydrogel in the presence of polyacid or polybases in an aqueous medium provides impossibility of transfer into a highly ionized state, due to what these areas, are areas of minimum conductivity. Maximum values of specific electric conductivity are observed at 33% hPMAA–67% hP2M5VP ratio at 48 h.

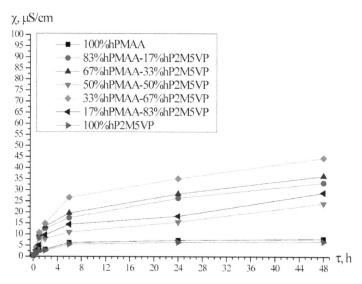

FIGURE 10.2 Dependence of specific electric conductivity of aqueous solution from time in presence of the intergel system hPMAA–hP2M5VP.

However, one cannot judge about the degree of mutual activation solely from the specific electric conductivity, because high values may indicate to the dominance of the dissociation of carboxyl groups over the proton association by the heteroatom of the poly-2-methyl-5-vinylpyridine (P2M5VP) hydrogel.

Figure 10.3 represents the dependence of the concentration of hydrogen ions in an aqueous medium from time in presence of the intergel system hPMAA–hP2M5VP. Minimum values of pH are observed in the presence of only acid hydrogel during all time of polymer interaction with the solution. Such phenomenon is due to the release of protons in the solution as a result of carboxyl group's electrolytical dissociation. Obtained results show that high cleavage rate of H^+ is observed before 2 h. Maximum values of pH are observed at 67% hPMAA–33% hP2M5VP at 48 h of hydrogels remote interaction in an aqueous medium.

So, the concentration of H^+ ions determines equilibrium in two processes:

1. Swelling rate, at which H^+ ions are released in solution as a result of –COOH groups dissociation.
2. Binding rate of H^+ ions by links of polybases.

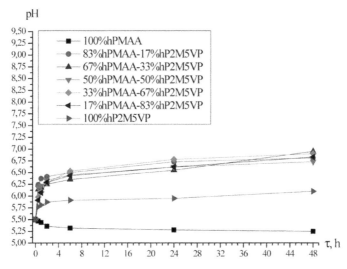

FIGURE 10.3 Dependence of pH of the aqueous solution from time in presence of the intergel system hPMAA–hP2M5VP.

Figure 10.4 shows that there is a swelling degree increase due to the long-range effect of the polymer hydrogels in the intergel system hPMAA–hP2M5VP.

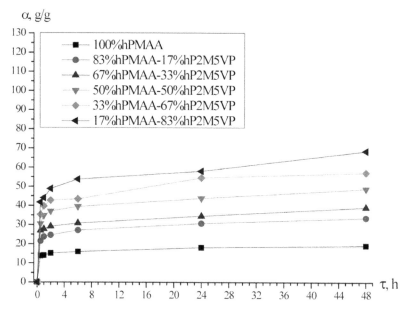

FIGURE 10.4 Dependence of hPMAA swelling degree in presence of hP2M5VP from time in an aqueous medium.

Minimum swelling of hPMAA is observed in the presence of individual hPMAA in aqueous solution during 48 h. Swelling of polyacid increases proportionally to increase of the share of polybases in the solution. Maximum values of the swelling degree of PMMA are seen at 48 h of remote interaction of the hydrogels at ratio 17% hPMAA:83% hP2M5VP. A significant increase in the swelling of polyacid is true for all ratios during 6 h. Further swelling occurs more slightly.

Figure 10.5 characterizes change of the swelling degree of hP2M5VP in an aqueous medium in the presence of hPMAA. With polyacid share increase in solution, there is additional swelling of the basic hydrogel of hP2M5VP.

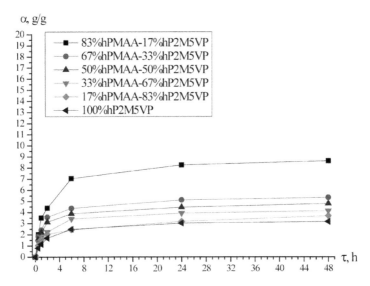

FIGURE 10.5   Dependence of hP2M5VP swelling degree in presence of hPMAA from time in an aqueous medium.

This phenomenon points to additional ionization of links of the polybases in the result of mutual activation of hydrogels of polymethacrylic acid (PMAA) and P2M5VP. It should be noted that polybases share increase provides an increase of protons association degree of heteroatom of polyvinylpyridine, what provides a shift of equilibrium to the right in the reaction of dissociation of carboxyl groups. Minimum values of the swelling degree of the polybases are observed in the presence of only hP2M5VP during all the time of interaction of the polymer with water. Maximum values of the parameter are observed at ratio 83% hPMAA–17% hP2M5VP at 48 h of hydrogels remote interaction.

Mutual activation of polymer hydrogels provides a significant increase in the swelling degree of hP2M5VP. As known, hP2M5VP is a weak polybases, consequently, has low values of swelling degree in an aqueous medium. A significant increase in the swelling of hP2M5VP points to high ionization degree of vinylpyridines links. This may be due to binding of the proton, which was cleaved from the carboxyl group, by nitrogen atoms of polybases.

Obtained data shows that hydrogels of PMAA and P2M5VP undergo remote interaction in an aqueous medium. The result of such interaction is the formation of same-charged groups with uncompensated counterions.

10.2.7 SORPTION OF LANTHANUM IONS BY THE INTERGEL SYSTEM hPMAA–hP2M5VP

Lanthanum nitrate is present in solution in a dissociated state. Dissociation of lanthanum nitrate occurs in three stages. Dissociation constant of the first stage is much higher than the constant of second and third. Due to this fact, sorption of dissociated ions in the intergel system occurs according to different mechanisms. Nitrate of lanthanum formed in the first stage of dissociation is binded by an ionic mechanism. In the result of dissociation of lanthanum nitrate by second and third stages, there is an increase of ionic pairs and molecular forms concentration in solution, what, in turn, provides their binding by coordination mechanism.

In the presence of the intergel system in the salt solution, there is an occurrence of the following reaction:

1. Dissociation of lanthanum nitrate along with dissociation of carboxyl groups.
2. Mutual activation of the hydrogels due to protons association by polybases.
3. Sorption of lanthanum ions.

These reactions impact on electrochemical equilibrium, and dominance of any of them will provide changes of specific electric conductivity and pH.

In a solution of lanthanum nitrate in presence of abovementioned intergel system, in the beginning, there is ionization of the polymer hydrogels similarly to the mutual activation in an aqueous medium; further ionization occurs due to the formation of coordination bonds with lanthanum ions. Due to sorption of lanthanum by coordination mechanism, the polymers do not have the same charged on internode links, which provides folding of the macromolecules and swelling decrease.

Extraction of lanthanum ions from the solution by the intergel system hPMAA–hP2M5VP is accompanied with a decrease of specific electric conductivity (Fig. 10.6). The significant decrease of electric conductivity occurs during 30 min after start of contact of the intergel system with lanthanum nitrate solution. Character of decrease of electric conductivity at all ratios of hydrogels is different. At a ratio 83% hPMAA–17% hP2M5VP after 6 h, decrease of the parameter is not very intensive. This indicates to not high degree of ionization of the polymer structures during their remote interaction. Minimum values of electric conductivity in the solution of lanthanum nitrate are observed at a ratio 50% hPMAA–50% hP2M5VP at 48 h. During mutual activation of the polymer hydrogels, there is a release of protons in the solution, however, electric conductivity decreases, which points to the fact of the metal sorption by the polymers.

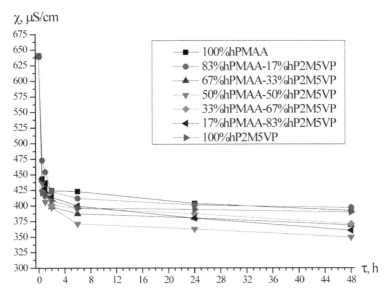

FIGURE 10.6 Dependence of specific electric conductivity of lanthanum nitrate from time in presence of the intergel system hPMAA–hP2M5VP.

Dependence of pH of lanthanum nitrate solution from time in presence of the intergel system hPMAA–hP2M5VP is presented in Figure 10.7.

As seen from Figure 10.7, minimum values of pH are observed in presence of only polyacid. It allows to make a conclusion that there is a release of protons in the solution due to the shift of the equilibrium to the right.

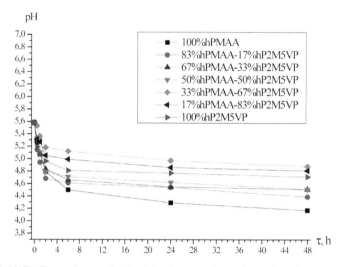

FIGURE 10.7   Dependence of pH of lanthanum nitrate from time in presence of the intergel system hPMAA–hP2M5VP.

Dependence of the swelling degree of the hPMAA from time in the presence of the hydrogel of hP2M5VP is presented in Figure 10.8. Maximum swelling of hPMAA occurs at 30 min of the interaction of the intergel system with the solution at ratio 17% hPMAA–83% hP2M5VP.

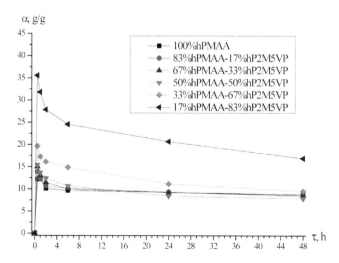

FIGURE 10.8   Dependence of hPMAA swelling degree in presence of hP2M5VP from time in lanthanum nitrate.

Dependence of the swelling degree of basic hydrogel of hP2M5VP from time in presence of hPMAA is shown in Figure 10.9. Change of swelling of the polybases is similar to polyacid's swelling—increase of concentration of second hydrogel (PMMA) provides increase of swelling degree. Maximum values of the swelling degree of hP2M5VP are reached at 30 min of remote interaction of the hydrogels in salt solution at ratio 50% hPMAA–50% hP2M5VP. Impossibility of transition into a highly ionized state of the polybases in case of presence of only hP2M5VP provides absence of additional activation of links, which promotes unfolding of macromolecular globe. As a result, minimum values of the swelling degree of the basic hydrogel.

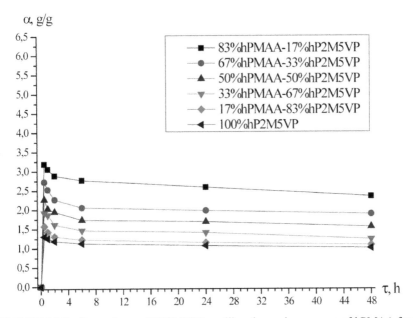

FIGURE 10.9 Dependence of hP2M5VP swelling degree in presence of hPMAA from time in lanthanum nitrate.

Figure 10.10 shows dependence of lanthanum ions extraction degree of the intergel system hPMAA–hP2M5VP from hydrogels' molar ratios in time. With time, extraction degree increases.

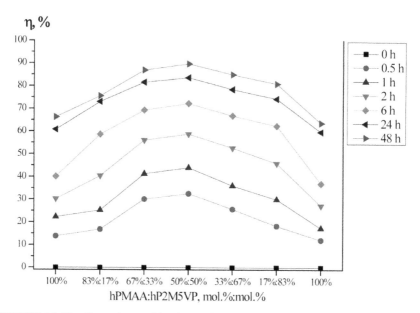

FIGURE 10.10 Dependence of lanthanum ions extraction degree of the intergel system hPMAA–hP2M5VP from hydrogels' molar ratios in time.

Extraction degree of lanthanum of individual hydrogels of PMAA and P2M5VP is not high. hPMAA sorbes 66.28% of lanthanum, hydrogel of hP2M5VP—63.65%. Increase of sorption ability of the initial hydrogels in the intergel system is due to their transition into a highly ionized state in the result of hydrogels mutual activation during their remote interaction in the intergel system. The highest values of sorption degree are reached at 67% hPMAA–33% hP2M5VP and 50% hPMAA–50% hP2M5VP ratios. A maximum amount of lanthanum is extracted at 50% hPMAA–50% hP2M5VP hydrogels ratio, extraction degree of lanthanum ions is 89.65%.

Dependence of polymer chain binding degree (in relation to lanthanum ions) of the intergel system hPMAA–hP2M5VP from time is presented in Figure 10.11.

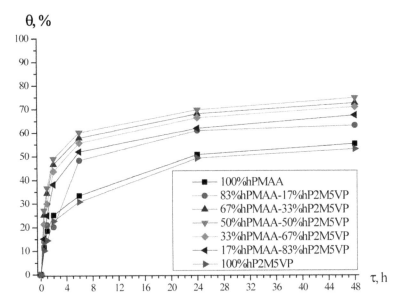

FIGURE 10.11 Dependence of polymer chain binding degree of the intergel system hPMAA–hP2M5VP from hydrogels' molar ratios in time.

Individual hydrogels of PMAA and P2M5VP have not very high binding degree: 55.17% for hPMAA and 53.00% for hP2M5VP. In result of mutual activation of polymer hPMAA and hP2M5VP, significant changes in their sorption properties occur. The highest values of polymer chain binding degree are observed at 50% hPMAA–50% hP2M5VP, the binding degree is 74.67%.

Dependence of effective dynamic exchange capacity of the intergel system hPMAA–hP2M5VP from hydrogels' molar ratios in time is presented in Figure 10.12. As seen from obtained results, the intergel system based on rare crosslinked polymer hPMAA and hP2M5VP has significantly higher values of exchange capacity comparatively with individual hydrogels. It should be noted that two polymers have bulk methyl substituents and their ionization occur slow. High values of the parameter are seen at ratios of 67% hPMAA–33% hP2M5VP and 50% hPMAA:50% hP2M5VP. Not sufficiently high values of exchange capacity are observed at ratios 83% hPMAA–17% hP2M5VP and 17% hPMAA–83% hP2M5VP due to the not very high degree of ionization of the polymers.

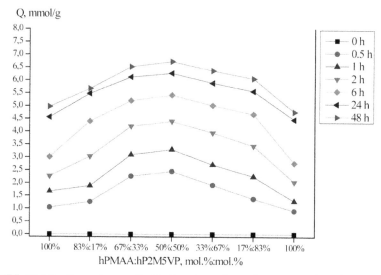

FIGURE 10.12 Dependence of effective dynamic exchange capacity of the intergel system hPMAA–hP2M5VP from hydrogels' molar ratios in time.

Obtained results point to the fact that the transition of polymer hydrogels in the intergel systems into a highly ionized state is characterized by a significant increase (up to 30%) of effective dynamic exchange capacity comparatively with initial hydrogels.

In Table 10.1, sorption parameters (for lanthanum sorption) of individual hydrogels of PMAA and P2M5VP and the intergel system on their basis are presented.

TABLE 10.1 Sorption Parameters of Individual Hydrogels and the Intergel System During Lanthanum Sorption.

Individual hydrogel/intergel system	hPMAA	hP2M5VP	50% hPMAA–50% hP2M5VP
Lanthanum extraction degree (%)	66.28	63.65	89.65
Polymer chain binding degree (%)	55.17	53.00	74.67
Effective dynamic exchange capacity (mmol/g)	4.97	4.77	6.72

As seen from the table, there is a significant increase in binding ability of the initial polymers in intergel pairs, what is due to their transition into a highly ionized state. It should be also noted that there are differences in behavior of the intergel system hPMAA–hP2M5VP in lanthanum nitrate comparatively with activation of the polymers in an aqueous medium. There is a decrease of specific electric conductivity, increase of hydrogen ions concentration, and the increase of the swelling degree of both hydrogels with further decrease of this parameter.

10.2.9 SORPTION OF CERIUM IONS BY THE INTERGEL SYSTEM hPMAA–hP2M5VP

Dependence of cerium ions extraction degree of the intergel system from hydrogels' molar ratios in time is presented in Figure 10.13.

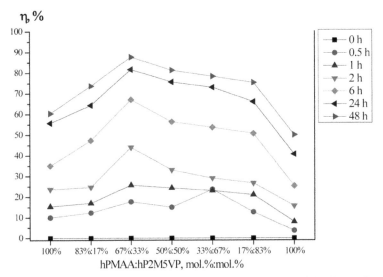

FIGURE 10.13 Dependence of cerium ions extraction degree of the intergel system hPMAA–hP2M5VP from hydrogels' molar ratios in time.

As seen from the figure, individual hydrogels of PMAA and P2M5VP have not very high degree of cerium ions extraction (60.33% for hPMAA and 50.00% for hP2M5VP). Maximum values of cerium sorption are observed at hydrogels ratio 67% hPMAA–33% hP2M5VP, sorption degree is 87.67%.

Dependencies of polymer chain binding degree (in relation to cerium ions) of the intergel system hPMAA–hP2M5VP from time are shown in Figure 10.14. Similarly to other parameters, polymer chain binding degree is in great influence from state, in which polymer hydrogel presents in solution. In other words, hydrogels in intergel pairs are in a highly ionized state due to their mutual activation in the result of their remote interaction. As seen from the obtained data, the maximum values of binding degree are observed at hydrogels ratio 67% hPMAA–33% hP2M5VP, it is 72.72%. Also, high values are in intergel pair 50% hPMAA–50% hP2M5VP, what, in turn, points to high ionization degree of the hydrogels in the intergel pair. Not sufficiently high values of polymer chain binding degree at hydrogels ratios 83% hPMAA–17% hP2M5VP and 17% hPMAA–83% hP2M5VP are the consequence of not very high ionization degree of the polymer macromolecules. Interaction of the individual hydrogels of PMAA and P2M5VP with cerium nitrate shows that equilibrium in the solution is reached rather fast, a subsequence of what is not high values of polymer chain binding degree. Binding degree is 50.05% for hPMAA and 41.47% for hP2M5VP, respectively.

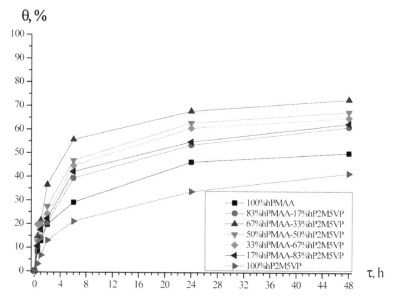

FIGURE 10.14 Dependence of polymer chain binding degree of the intergel system hPMAA–hP2M5VP from hydrogels' molar ratios in time.

Figure 10.15 reflects dependence of effective dynamic exchange capacity (in relation to cerium ions) of the intergel system hPMAA–hP2M5VP from hydrogels' molar ratios in time. The results show that the maximum values of the parameter are observed at hydrogels ratio 67% hPMAA–33% hP2M5VP at 48 h. High values of exchange capacity are also observed at 50% hPMAA–50% hP2M5VP and 33% hPMAA–67% hP2M5VP ratios. The minimum values of effective dynamic exchange capacity are observed in the presence of individual polymer hydrogels of PMAA and P2M5VP. This is the result of the absence of the phenomenon of mutual activation.

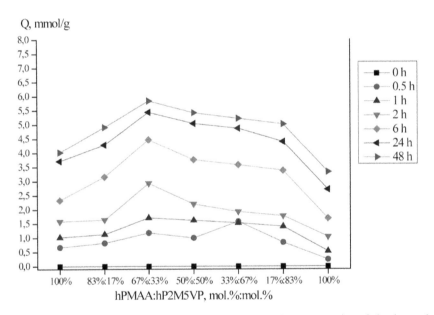

FIGURE 10.15 Dependence of effective dynamic exchange capacity of the intergel system hPMAA–hP2M5VP from hydrogels' molar ratios in time.

In Table 10.2, sorption parameters (for cerium sorption) of individual hydrogels of PMAA and P2M5VP and the intergel system on their basis are presented.

TABLE 10.2 Sorption Parameters of Individual Hydrogels and the Intergel System During Cerium Sorption.

Individual hydrogel/intergel system	hPMAA	hP2M5VP	67% hPMAA–33% hP2M5VP
Cerium extraction degree (%)	60.33	50.00	87.67
Polymer chain binding degree (%)	50.05	41.47	72.72
Effective dynamic exchange capacity (mmol/g)	4.02	3.33	5.84

As seen from Table 10.2, the sorption parameters of the intergel system hPMAA–hP2M5VP is almost on 25–30% higher in comparison with individual hydrogels of PMAA and P2M5VP. Remote interaction of the hydrogels in the intergel system provides a significant increase in cerium ions extraction degree, polymer chain binding degree, and effective dynamic exchange capacity.

10.2.10   SELECTIVE SORPTION OF LANTHANUM IONS BY THE INTERGEL SYSTEM hPMAA–hP2M5VP

For selective extraction of lanthanum ions from solution, which contains ions of lanthanum and cerium, the ratio 50% hPMAA–50% hP2M5VP was taken due to the fact that maximum sorption of lanthanum occurs at this ratio.

Figure 10.16 represents dependencies of lanthanum and cerium ions extraction degrees of the intergel system 50% hPMAA–50% hP2M5VP from time. High level of mutual activation in the intergel pair during 2 h provides a significant increase in lanthanum and cerium ions extraction. At this time of hydrogels, remote interaction in the solution 29.25% of lanthanum and 7.56% of cerium is extracted from the solution. Further, there is an occurrence of extraction degree of both the metals. At 24 h of the hydrogels interaction, lanthanum ions sorption degree is 52.97% and 18.46% for cerium ions. It should be noted that increase up to 48 h occurs very slightly, which indicates that the system is reaching the equilibrium state. At 48 h, 65.40% of lanthanum is sorbed and 24.38% of cerium is sorbed.

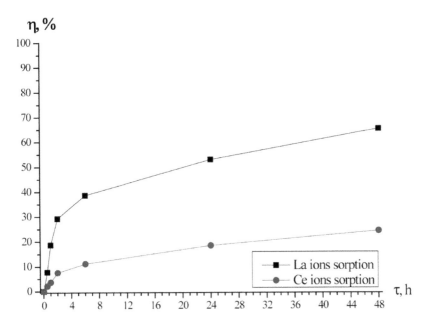

FIGURE 10.16 Dependence of lanthanum and cerium ions extraction degree of the intergel system 50% hPMAA–50% hP2M5VP from time.

Dependence of polymer chain binding degree (in relation to lanthanum and cerium ions) of the intergel system 50% hPMAA–50% hP2M5VP from time is presented in Figure 10.17. As seen from the obtained data, a sharp increase of polymer chain binding degree of the intergel system is observed during the first 2h, binding degree of lanthanum is 24.26% and of cerium is 6.27%. Further interaction of the polymer hydrogels in the intergel pair provides a subsequent increase of binding degree. Maximum values of binding degree are observed at 48 h of remote interaction, wherein 54.25% of lanthanum and 20.22% of cerium is bind.

Figure 10.18 presents dependence of effective dynamic exchange capacity of the intergel system 50% hPMAA–50% hP2M5VP from time. There is an increase in the parameter with time.

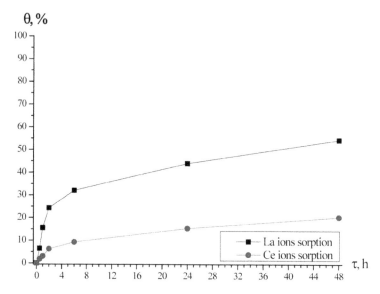

FIGURE 10.17   Dependence of polymer chain binding degree of the intergel system 50% hPMAA–50% hP2M5VP from time.

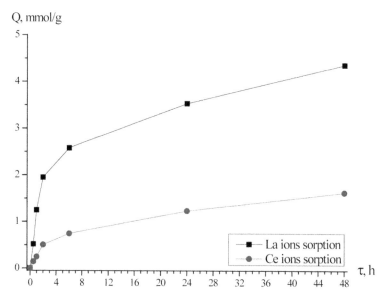

FIGURE 10.18   Dependence of effective dynamic exchange capacity of the intergel system 50% hPMAA–50% hP2M5VP from time.

Effective dynamic exchange capacity in relation to lanthanum is in 2.5 times higher compared with cerium. These data evidence that the intergel system 50% hPMAA–50% hP2M5VP shows selectivity to lanthanum ions during its sorption from solution, which contains lanthanum and cerium ions.

In Table 10.3, the values of the sorption parameters (extraction degree, polymer chain binding degree, and effective dynamic exchange capacity) of the intergel system 50% hPMAA–50% hP2M5VP at selective sorption of lanthanum ions are presented. It should be noted that these values are the results of measurements at 48 h.

TABLE 10.3   Sorption Parameters of the Intergel System 50% hPMAA—50% hP2M5VP at Lanthanum Selective Sorption.

Intergel system	50% hPMAA–50% hP2M5VP
Lanthanum extraction degree (%)	65.40
Cerium extraction degree (%)	24.38
Polymer chain binding degree (in relation to lanthanum) (%)	54.25
Polymer chain binding degree (in relation to lanthanum) (%)	20.22
Effective dynamic exchange capacity (in relation to lanthanum) (mol/g)	4.36
Effective dynamic exchange capacity (in relation to lanthanum) (mmol/g)	1.63

10.3   CONCLUSION

1.   Mutual activation of PMAA and P2M5VP hydrogels in the intergel system on their basis in an aqueous medium provides a significant increase in electrochemical properties of polymer macromolecules. Specific electric conductivity increases, while the concentration of hydrogen ions decreases.

2.   The result of remote interaction of polymer hydrogels of PMAA and P2M5VP in the water medium is a sharp increase of swelling, what is due to the formation of same charges of internode links of polymer chains, what, in turn, provides unfolding of macromolecular globe and increase of swelling degree.

3. Sorption of lanthanum ions by the intergel system hPMAA–hP2M5VP is accompanied by radically different changes of electrochemical properties of polymer hydrogels. There is a decrease of specific electric conductivity and an increase of protons concentration in the solution. Also, it should be noted that in the process of lanthanum sorption amount of same-charged groups on internode links of polymer chains provides folding of polymer globe and decrease of swelling degree.

4. All intergel pairs in the intergel system hPMAA–hP2M5VP have much higher values of lanthanum ions extraction degree comparatively with initial hydrogels of PMAA and P2M5VP. Wherein maximum sorption of lanthanum is observed at 50% hPMAA–50% hP2M5VP ratio, sorption degree is 89.65%. Polymer chain binding degree (in relation to lanthanum ions) has maximum values at hydrogels ratio 50% hPMAA–50% hP2M5VP, the value of the binding degree is 74.67%.

5. The intergel system hPMAA–hP2M5VP have maximum values of cerium ions extraction degree at 67% hPMAA–33% hP2M5VP (sorption degree is 87.67%). Polymer chain binding degree at this ratio also has the highest values, it is 72.72%.

6. Selective sorption of lanthanum ions by the intergel system 50% hPMAA–50% hP2M5VP indicates that selectivity of the intergel system is manifested to lanthanum ions. Extraction degree of lanthanum ions is 65.40%, and for cerium ions, it is 24.38%. Polymer chain binding degree is in relation to lanthanum ions—54.25%, in relation to cerium ions—20.22%.

ACKNOWLEDGMENTS

The work was financially supported (the work was done due to the grant funding of two Projects: AP05131302 and AP05131451) by the Committee of Science of Ministry of Education and Science of the Republic of Kazakhstan.

KEYWORDS

- polymethacrylic acid
- poly-2-methyl-5-vinylpyridine
- remote interaction
- La^{3+} ions
- Ce^{3+} ions
- sorption

REFERENCES

1. Alimbekova, B. T.; Korganbayeva, Zh. K.; Himersen, H.; Kondaurov, R. G.; Jumadilov, T. K. Features of Polymethacrylic Acid and Poly-2-methyl-5-vinylpyridine Hydrogels Remote Interaction in an Aqueous Medium. *J. Chem. Chem. Eng.* **2014**, *3*, 265–269.

2. Alimbekova, B.; Erzhet, B.; Korganbayeva, Zh.; Himersen, H.; Kaldaeva, S.; Kondaurov, R.; Jumadilov, T. Electrochemical and Conformational Properties of Intergel Systems Based on the Crosslinked Polyacrylic Acid and Vinylpyridines. In: *Proceedings of VII International Scientific-Technical Conference "Advance in Petroleum and Gas Industry and Petrochemistry"* (APGIP-7), Lviv, Ukraine, May 2014; p 64.

3. Jumadilov, T. K.; Himersen, H.; Kaldayeva, S. S.; Kondaurov, R. G. Features of Electrochemical and Conformational Behavior of Intergel System Based on Polyacrylic Acid and Poly-4-vinylpyridine Hydrogels in an Aqueous Medium. *J. Mater. Sci. Eng., B* **2014**, *4*, 147–151.

4. Jumadilov, T. K.; Abilov, Zh. A.; Kaldayeva, S. S.; Himersen, H.; Kondaurov, R. G. Ionic Equilibrium and Conformational State in Intergel System Based on Polyacrylic Acid and Poly-4-vinylpyridine Hydrogels. *J. Chem. Eng. Chem. Res.* **2014**, *1*, 253–261.

5. Jumadilov, T. K.; Abilov, Zh. A.; Kondaurov, R. G.; Eskalieva, G. K. Mutual Activation of Hydrogels of Polymethacrylic Acid and Poly-2-methyl-5-vinylpyridine. *Chem. J. Kazakhstan* **2015**, *2*, 75–79.

6. Jumadilov, T.; Abilov, Zh.; Kondaurov, R.; Himersen, H.; Yeskalieva, G.; Akylbekova, M.; Akimov, A. Influence of Hydrogels Initial State on Their Electrochemical and Volume-Gravimetric Properties in Intergel System Polyacrylic Acid Hydrogel and Poly-4-vinylpyridine Hydrogel. *J. Chem. Chem. Technol.* **2015**, *4*, 459–462.

7. Jumadilov, T.; Kaldayeva, S.; Kondaurov, R.; Erzhan, B.; Erzhet, B. Mutual Activation and High Selectivity of Polymeric Structures in Intergel Systems. In *High Performance Polymers for Engineering Based Composites*; Jumadilov, T., Omari, V., Mukbaniani, M., Abadie, J. M., Tatrishvilli, T.; CRC Press: Boca Raton, FL, 2015; pp 111–119.

8. Petruhin, O. M. Methodology of Physico-chemical Methods of Analysis. *M.: Chemistry* **1987**, 77–80.

CHAPTER 11

SYNTHESIS AND PROPERTIES OF GALLIC ACID-GRAFTED PHLOROGLUCINOL/ FORMALDEHYDE COMPOSITE

AMJAD MUMTAZ KHAN* and YAHIYA KADAF MANEA

Department of Chemistry, Analytical Research Lab, Aligarh Muslim University, Aligarh 202002, Uttar Pradesh, India

*Corresponding author. E-mail: amjad.mt.khan@gmail.com

ABSTRACT

In this study, phloroglucinol–formaldehyde composite (PFC)-grafted gallic acid as a functional species has been synthesized. The PFC was characterized using Fourier transform infrared spectroscopy, X-ray diffraction, thermal analysis (TGA–DTA), scanning electron microscopy and ^{13}C NMR techniques. The material shows high chemical stability and moderate thermal stability. The environmental utility has been achieved on the basis of selective removal of Hg^{2+}, Pb^{2+}, Cu^{2+}, and Zn^{2+} ions from dilute aqueous solutions. The distribution coefficient of PFC composite toward metal ions has been studied in varying concentration of tartaric acid as well as buffer solutions at different pH.

11.1 INTRODUCTION

Phenols belong to the family of aromatic compounds with hydroxyl group bonded directly to an aromatic nucleus. They differ from alcohols in that they behave like weak acids and dissolve readily in aqueous sodium hydroxide but are insoluble in aqueous sodium carbonate. Phenols are colorless solids with exception of some liquid alkyl phenols. The chemistry of phenolic resins involves a variety of key factors which are critical in the design of desired phenolic resin.[1] These low-to-medium molecular weight material can be viewed as "reactive intermediates" which can be cured or can undergo many transformation reactions by appending new reactive groups to phenolic hydroxyl substituent such as epoxy or form a new ring structure. Phenolic resins are obtained by step growth polymerization.[2] Chelating resins have been widely utilized for the removal of the undesired metal ions from aqueous solution.[3–6] Removal and separation of metal ions in aqueous solution play an important role for the analysis of wastewaters, industrial and geological samples as well as for environmental remediation.[7] Toxic heavy metal ions are well known about their serious harm to human health. If the toxic metal ions can be recovered, the energy and material requirements of the wastewater treatment process can be simpler. Several techniques have been developed for the treatment of such waste streams. Among these, the ion exchange technique is the most common and effective method for the treatment and identification of industrial wastes.[8] The use of chelating resins for separation and removal of ions by the method of choice due to its high separation efficiency, good reproducibility of retention parameters and simplicity of operation.[9] Therefore, the development of high-performance chelating resins for removing heavy metal ions from aqueous solution is considered as a research priority in the environmental field.[10] Metal ion recovery proves its applications related to hydrometallurgy and analytical methods has created an increased interest in the development of chelating ion exchangers. Chemistry regarding the use of ion exchange and chelating polymers for separation and preconcentration of trace metal ions is studied.[11,12] A new chelating sorbent based on pyrazolone-containing amines immobilized on styrene divinylbenzene copolymer have been reported selective for Au^{III}, Pb^{II}, and Ag^{I}.[13] The chelating resins DowexA1 and its purified form, chelex-100, ion fauns the imidoacetic functional group.[14] The affinity of polyproduct of a metal sulfide for Duolite GT-73 chelating resin was investigated along with the pH effect of the eluents on the affinity for ions.[15] The sorption of Cu, Hg, Pb, and Zn ions on thiol (–SH)-based chelating polymeric resin.[16] Stannic silicate has reacted with complexions (xylenol range, EBT, 1–10 phenanthroline) for the use as chelating ion exchange.[17] Some new amphoteric ion exchange resins were synthesized by condensing catechol,

8-hydroxy-quinoline, and hydroquinone. The present study is focused on the synthesis of phenol–formaldehyde-based chelating resin with gallic acid as the functional species. The phenol-based polymer can also be used in the synthesis of nanoporous carbon aerogels and xerogels.[18,19] It can further be explored for the determination of metal ions in synthetic as well as in industrial samples.

11.2   EXPERIMENTAL

11.2.1   REAGENTS AND MATERIALS

All reagents used for the synthesis were obtained from E-Merck and S.D. Fine Chemicals India.

11.2.2   SYNTHESIS OF PHLOROGLUCINOL–FORMALDEHYDE COMPOSITE

A known volume of formaldehyde (30 mL) is taken and added to phenol (phloroglucinol, 20 mL) in a flask. Afterward to the above mixture, 10 mL of gallic acid is added. To the above mixture solution, 10 mL of sulfuric acid is added (as a catalyst). The whole ion mixture is heated on a magnetic stirrer for 3 h maintaining the temperature between 60°C and 70°C. Then it is again heated for 4 h at a higher temperature between 80°C and 90°C. Finally, it is cooled, filtered on the suction pump, washed with DMW to remove excess of reagents, dried in an electrical oven at 50°C and stored in air-tight desiccators.

11.2.2.1   REACTION SCHEME

HCHO (0.3 M) + phloroglucinol (0.3 M) + gallic acid (0.03 M) + H_2SO_4 (2 M)
30 mL   20 mL   10 mL   10 mL

SCHEME 11.1   Formation of thick precipitates occurs instantaneously.

11.2.3   PHYSICOCHEMICAL PROPERTIES

The chemical stability of the composite was examined in mineral acids, polar as well as ion nonpolar solvents. The material exhibited good stability in all solvents. The chemical composition of material carried out by elemental analysis by using CHN technique.

11.2.4   INSTRUMENTAL CHARACTERIZATION

The Fourier-transform infrared (FTIR) spectrum of composite dried at 40°C was taken by KBr disc method at room temperature using Perkin-Elmer, USA, Spectrum-BX spectronic FTIR. A 10-mg finally powdered polymer was thoroughly mixed with 100 mg of KBr and grounded to a fine powder. A transparent disc was formed by applying a pressure of 80,000 psi (1 psi = 6894.76) in a moisture-free atmosphere. The IR spectrum was recorded between 400 and 4000 cm^{-1}. X-ray diffraction (XRD) analysis of material was carried out using manganese-filtered and Cu $K\alpha$ radiation (λ = 1.5418) was used. The instrument X-ray diffracts meter-Philips (Holland); model PW-1148/89 is equipped with a graphite monochromator operating at 40 kV and 30 mA to perform XRD analysis. The X-ray studies were performed between 0° and 60° 2θ while the speed of the recorder was maintained at 10 mm/s 2θ. TGA–DTA studies were carried out by heating the sample 1 up to 800°C at a constant rate (~20°C min^{-1}) in a nitrogen atmosphere (flow rate of 35 mL min^{-1}) using Perkin Elmer instrument. Electron microscopic recording of the composite was carried by using a scanning electron microscope JOEL (Japan) JSM-6510 with an accelerating voltage of 15 kV.

11.2.5   EFFECT OF pH ON METAL ION EXCHANGE CAPACITY

The effect of pH on the metal ion exchange capacity of the composite was studied. Different sets of accurately weighed (0.5 g) dry samples were equilibrated for 3 h with buffer solution so that the composite attained desired pH value. After 3 h, buffer solutions were decanted, and 50 mL of 0.5 mg metal ion solutions of range pH from 3 to 7 were added. Metal ion solutions were equilibrated for 24 h with intermittent shaking at room temperature. After 24 h, the solutions were filtered to separate the resin and solution. After filtration, the pH was measured, and it was found that pH remains stable throughout the experiment (±0.1). The method was followed throughout the study to calculate the ion exchange capacity of the resin as

$$\text{Exchange capacity} = \frac{\text{Initial molarity of the metal ion} - \text{Remaining molarity of the metal ion}}{\text{Atomic molar mass of the metal} \times \text{Weight of the sample}}$$

11.2.6 EFFECT OF METAL ION CONCENTRATION ON EXCHANGE CAPACITY

To study the effect of metal ion concentration by the synthesized composite, the accurately weighed (0.5 g) dry resin was equilibrated with buffer at desired pH values (pH value of highest exchange) for 3 h and then buffer solutions were decanted and metal ion solutions (50 mL) of different molar concentration 10–50 ppm of the same pH value were added and equilibrated at room temperature for 24 h with intermittent shaking. After 24 h, metal ion solutions were filtered and the unchelated metal ions were estimated.

11.2.7 DISTRIBUTION STUDIES FOR METAL IONS

Measurement of the distribution coefficient of metal ions over a wide range of condition is the best method to avoid choosing elution conditions for separation columns by a strictly trial and error process. The distribution coefficient (K_d) for Hg^{2+}, Pb^{2+}, Zn^{2+}, and Cu^{2+} ions was determined in the presence of citric acid. A dry resin sample weighed exactly (0.5 g) was put in a 100-mL conical flask each containing 30 mL of 2×10^{-4} M of metal ion in the presence of citric acid. The pH of the solution was adjusted to the desired value using acetate buffer and the resin has been equilibrated for 24 h. After 24 h, the solutions were decanted and unabsorbed metal ions were estimated.[19] Distribution coefficient values were calculated by using the following relationship:

$$K_d = \frac{I - F}{F} \times \frac{V}{M} \left(\text{mL g}^{-1} \right)$$

where I is the initial amount of the metal ion in the solution phase, F is the final amount of metal ion in the solution phase after treatment with the exchanger, V is the volume of the solution (mL), and M is the amount of ion exchanger taken (g).

11.3 RESULTS AND DISCUSSIONS

Table 11.1 shows the synthesis of the composite under varying experimental conditions. It was finally concluded on the basis of product yield and thermal studies; sample 1c was finally selected for further studies. FTIR spectrum of the material is shown in Figure 11.1. A sharp and broad peak at 3400 cm^{-1} is the characteristic of –OH group. The bands at 1460 and 1265 cm^{-1} indicate to the formation of methylene and methylene ether bridges between the aromatic rings. These peaks correspond respectively to the C–H bending and C–O stretching vibrations in C–O–C structures. The main band in the region of 1630 cm^{-1} is due to the C=C stretching in the aromatic rings, while the bands at around 1114 and 1340 cm^{-1} may also be attributed to C–OH stretching and deformation in phenolic groups.[20] The ^{13}C NMR spectrum provides useful information about the nature of the carbon present in the sample. The spectrum shows the corresponding peaks at 114.4, 153.5, and 127.8 ppm with respect to C_1–C_3 of the aromatic ring. The peak appeared at 41.12 ppm is assigned to the Ar–CH$_2$ in the resin. The appeared shift at 22.54 ppm is assigned to –CH$_2$–CH$_2$– linkage in the composite (as shown in Fig. 11.2). The peak appeared at 182.4 ppm is indicates to C=O group in the material. The XRD patterns of the compound provide the information about the crystalline or amorphous nature of the compound or crystalline and amorphous regions that may co-exist in the same compound.[21,22] Figure 11.3 shows the X-ray powder diffraction pattern of the polymer. The spectrum of the polymer shows the presence of mixed amorphous and crystalline phases indicating its semi-crystalline nature. Table 11.2 shows the chemical stability of the polymer. It can be readily concluded from the data that the polymer shows high chemical stability in most of the solvents studied. The morphology of the resin sample was carried out by scanning electron micrographs, which is shown in Figure 11.4. The morphology of the resin shows a fringed model of the crystalline–amorphous structure. Figure 11.5 shows the thermal analysis (TGA–DTA) curve of the polymer in the nitrogen atmosphere.

TABLE 11.1 Synthesis Conditions.

Set	HCHO		Phenol		Gallic acid		H$_2$SO$_4$		Yield (mg)
	Conc. (M)	Volume (mL)	Conc. (M)	Volume (mL)	Conc. (M)	Volume (mL)	Conc. (M)	Volume (mL)	
1a	0.1	25	0.1	10	0.01	10	2	10	25
	0.2	25	0.2	10	0.02	10	2	10	30
1c	**0.3**	**25**	**0.3**	**10**	**0.03**	**10**	**2**	**10**	100
	0.4	25	0.4	10	0.04	10	2	10	40
	0.5	25	0.5	10	0.05	10	2	10	50
	0.6	25	0.6	10	0.06	10	2	10	55
2a	0.1	25	0.1	10	0.01	10	0.1	20	60
	0.2	25	0.2	10	0.02	10	0.2	20	45
	0.3	25	0.3	10	0.03	10	0.3	20	30
	0.4	25	0.4	10	0.04	10	0.4	20	60
	0.5	25	0.5	10	0.05	10	0.5	20	70
	0.6	25	0.6	10	0.06	10	0.6	20	50
3a	0.1	30	0.1	20	0.01	10	0.1	20	55
	0.2	30	0.2	20	0.02	10	0.2	20	40
	0.3	30	0.3	20	0.03	10	0.3	20	60
	0.4	30	0.4	20	0.04	10	0.4	20	50
	0.5	30	0.5	20	0.05	10	0.5	20	50
	0.6	30	0.6	20	0.06	10	0.6	20	60
4a	0.1	30	0.1	10	0.01	20	0.1	10	40
	0.2	30	0.2	10	0.02	20	0.2	10	30
	0.3	30	0.3	10	0.03	20	0.3	10	55
	0.4	30	0.4	10	0.04	20	0.4	10	40
	0.5	30	0.5	10	0.05	20	0.5	10	50
	0.6	30	0.6	10	0.06	20	0.6	10	60

TABLE 11.2   Chemical Stability of Polymer in Various Solvents.

Sr. no.	Solvents	Remarks
1	DMW	Completely insoluble
2	Chloroform	Moderately soluble
3	Benzene	Sparingly soluble
4	CCl_4	Insoluble
5	Ethanol	Sparingly soluble
6	Methanol	Completely insoluble
7	DMF	Completely insoluble
8	Acetonitrile	Completely insoluble
9	Formamide	Completely insoluble
10	Pyridine	Partially soluble
11	Perchloric acid	Partially soluble
12	Acetone	Completely insoluble
13	1 M HCl	Completely insoluble

DMW, deminaralized water.

TABLE 11.3   Elemental Analysis and Physicochemical Properties of Phloroglucinol–Formaldehyde Resin.

Properties	Value
% C calculated (found)	53 (53.2)
% H calculated (found)	3 (3.1)
% O calculated (found)	43 (42.9)
% N calculated (found)	0 (0)
% Moisture	6 ($\pm$0.02)
% Solid	93.75 ($\pm$0.02)
True density	0.89 ($\pm$0.01) g/cm^3

TGA curve shows first weight loss at 100°C due to the removal of –OH groups and external water molecules. A slow and gradual weight loss occurs at 154°C. This may be due to the slow decomposition of the organic part. A sharp and intense weight loss occurs between 200°C and 300°C due to the fast decomposition of the organic part. After 300°C up to 750°C, the weight loss becomes negligible (6%) due to the formation of the final product. The synthesized material is thermally stable. DTA curve of the polymer shows the presence of the endothermic peak at 201°C due to dehydration and decomposition of product. The chemical composition of the composite was carried out by elemental analysis to determine the amount of carbon, hydrogen, and nitrogen. Physicochemical properties of the synthesized resin are presented in Table 11.3. The moisture content of the resin provides a measure of its water swelling capacity or its loading capacity. The water content depends on numerous factors, such that the composition of the composite matrix, the cross-linking degree, or the nature of the active groups and the ionic polymer form. The degree of cross-linking, a composite has an effect on the moisture content and mois-ture content of the resin, and therefore, has an effect on the selectivity. In a high moisture content of the resin, the active groups are more spaced apart. The water content percentage of synthetic resin, as shown in Table 11.3 is 7.25%. The % C, % H, and % N was calculated from the general formula $(C_{14}H_{10}O_8)$ of the repeating unit of the assumed structure (Scheme 11.1) of the synthesized composite. The chemical composition of the composite was carried out by elemental analysis to determine the amount of carbon, hydrogen, and nitrogen. The amount of carbon, hydrogen, and nitrogen was found to be 53%, 3%, and 0%. The amount of oxygen comes out to be 43% which is shown in Table 11.3. The surface of resin, as well as the solution chemistry of these metal ions, is pH dependent. Changes in pH are known to affect the adsorbent's surface charge and the adsorbate's degree of ionization and speciation.[23] The results of the exchange capacity depending on the pH for different metal ions are presented in Figure 11.6. The results show that the sorption metal ions is increased with increasing pH to a maximum value and thereafter decreased. Maximum sorption occurred Pb(II) at pH 5, Zn(II) at pH 4.5, Cu(II) at pH 6, and Hg(II) at pH 4.5. The order of selectivity for the metal ions is: Cu(II) > Zn(II) > Pb (II) > Hg(II). An increase in pH increases the negatively charged nature of the sorbent surface. This leads to an increase in the electrostatic attraction between positively charged metal ions and negatively charged sorbent and results in increased sorption of metal ions. At lower pH, the removal of metal ions is decreased due to the higher concentration of H^+ ions present in the reaction mixture which compete with the metal ions for the sorption sites at the surface. Meanwhile, the observed decrease in sorption capacity is due to the formation of insoluble metal ion hydroxides.[24] In the case of Hg(II), purely electrostatic factors are responsible. Due to the less deep pits, resin exhibits lower ion exchange capacity for Hg(II).

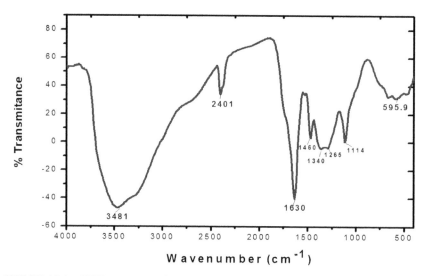

FIGURE 11.1   FTIR spectrum of the PFC composite.

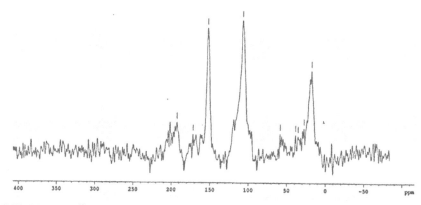

FIGURE 11.2 ^{13}C NMR spectrum of the PFC resin.

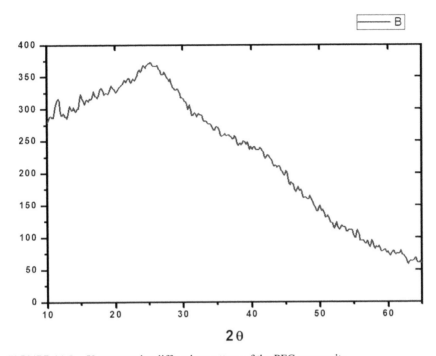

FIGURE 11.3   X-ray powder diffraction pattern of the PFC composite.

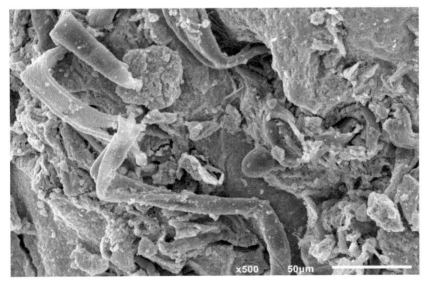

FIGURE 11.4   Scanning electron microscopy photographs of PFC composite at 500×.

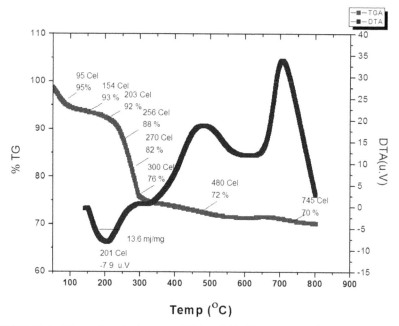

FIGURE 11.5 Thermal analysis curve (TGA–DTA) of PFC composite.

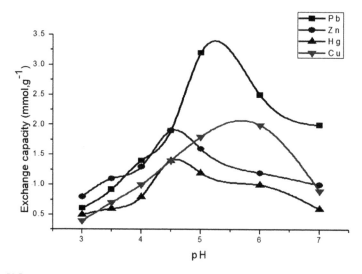

FIGURE 11.6 Effect of pH on cation exchange capacity.

Pb(II) has bigger hydrated radii, so it cannot easily penetrate to originate in the more crystalline region of the polymeric network. The examination of the data presented in Figure 11.7 reveals that the amount of metal ion adsorbed increases with the increase in the concentration of metal ions in solution until a maximum value and will remain constant upon further increase in metal ion concentration. A low concentration of metal ions, the available number of metal ions in the solution is low relative to the arrangement sites on the sorbent.[25] However, at higher concentrations, the sorption-available sites remain the same as more metal ions are available for sorption and subsequently sorption becomes almost constant then after.[26] For Hg(II), Cu(II), Zn(II), and Pb(II) after 300 ppm concentration, ion exchange attains limiting value. Distribution coefficient values (K_d) for different metal ions have been determined by the batch equilibration method. The K_d values have been investigated. Metal ions depending on pH in the presence of the tartaric acid and the results are shown in Table 11.4.

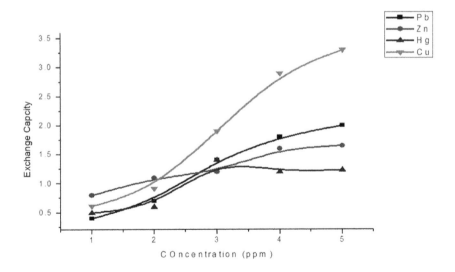

FIGURE 11.7   The effect of cation concentration on exchange capacity.

TABLE 11.4 K_d Values of Metal Ions at Various Citric Acid Concentrations and pH Values.

Metal ion	[Citric acid] (M)	K_d values at varying pH					
		3	3.5	4	4.5	5	6
Cu^{2+}	0.1	1181.6	771	611	517	401	320
	0.2	660.5	430	302	260	201	191
	0.3	221.7	180	110	98	80	70
	0.5	98.3	70	45.6	40	39	36
	1	85.6	40.5	37	35	32	30
Ni^{2+}	0.1	450	320	160	103	70	44
	0.2	330	212	140	94	67	41
	0.3	260	201	129	89	64	38
	0.5	190	146	102	73	56	35
	1	101	81	66	52	43	25
Hg^{2+}	0.1	401	309	221	136	83	54
	0.2	290	223	166	90	65	42
	0.3	146	125	88	63	52	38
	0.5	103	92	83	57	36	27
	1	88	74	61	46	32	21
Pb^{2+}	0.1	312	208	164	104	78	48
	0.2	204	145	92	72	65	45
	0.3	107	91	78	61	53	43
	0.5	89	75	62	55	39	28
	1	67	54	41	34	24	16

ACKNOWLEDGMENTS

The authors are thankful to the Chairman, Department of Chemistry, Aligarh Muslim University, Aligarh, India for providing necessary research facilities. Department of Physics, A.M.U., Aligarh is also acknowledged for recording XRD.

KEYWORDS

- composite
- phloroglucinol–formaldehyde
- column separations
- phloroglucinol–formaldehyde composite-grafted gallic acid
- phloroglucinol–formaldehyde resin

REFERENCES

1. Ellis, C. Synthetic Resins, *The Chemistry of Synthetic Resin*; Reinhold Publishers Company: New York, 1935.
2. *Coated Abrasive*, 2006. www.cumiabrasives.com.
3. Bisset, W.; Jacobs, H.; Koshti, N.; Stark, P.; Gopalan, A. Synthesis and Metal Ion Complexation Properties of a Novel Polyethyleneimine *N*-Methyl Hydroxamic Acid Water Soluble Polymer. *React. Funct. Polymer.* **2003**, *55* (2), 109–119.
4. Dikshit, D. *Orient J. Chem.* **2013**, *29* (1), 305–307.
5. Geckeler, K. E.; Volchek, K. Removal of Hazardous Substances from Water Using Ultrafiltration in Conjunction with Soluble Polymers. *Environ. Sci. Technol.* **1996**, *30* (3), 725–734.
6. Smith, B. F.; Gibson, R. R.; Jarvinen, G. D.; Robison, T. W.; Shroeder, N. C.; Stalnaker, N. D. Preconcentration of Low Levels of Americium and Plutonium from Waste Waters by Synthetic Water Soluble Metal-Binding Polymers with ultrafiltration. *J. Radioanal. Nucl. Chem.* **1998**, *234* (1–2), 225–229.
7. Shah, B. A.; Shah, A. V.; Shah, P. M. Analytical and Morphology Studies of Phthalic of Acid Formaldehyde Resorcinol as Chelating Resin. *Malaysian Polym. J.* **2009**, *4* (2), 1–12.
8. Khan, A. M.; Ganai, S. A.; Nabi, S. A. Synthesis of a Crystalline Organic Inorganic Composite Exchanger, Acrylamide Stannic Silicomolybdate: Binary and Quantitative Separation of Metal Ions. *Colloids Surf., A: Physicochem. Eng. Aspects* **2009**, *337*, 141–145.
9. Yirikoglu, H.; Gulfen, M. Separation and Recovery of Silver(I) Ions from Base Metal Ions by Melamine–Formaldehyde–Thiourea (MFT) Chelating Resin. *Separ. Sci. Technol.* **2008**, *4*, 376–388.
10. Wang, C.-C.; Wang, C.-C. Synthesis and Characterization of Chelating Resins with Amino Moieties and Application on Removal of Copper(II) from EDTA Complexes. *Inc. J. Appl. Polym. Sci.* **2005**, *97*, 2457–2468.
11. Schmuckler, G. *Talanta* **1965**, *10*, 745–751.
12. Kantipuly, C.; Kantragadda, S.; Chow, A.; Gesser, H. D. *Talanta* **1990**, *37*, 491.
13. Todoroba, O.; Ivanova, E.; Terebeniva, A.; Jordanov, N. *Talanta* **1989**, *36*, 817.
14. Rengen, K. *J. Radioanal., Nucl. Chem.* **1997**, *219* (2), 211–215.

15. Choi, B. S.; Sung, J. H. A Study on the Adsorption Behavior of Some Heavy Metals on Duolite GT-73 Chelating Resin. *Bull Korean. Chem. Soc.* **2000,** *21* (5), 538–540.

16. Saha, B.; Iglesias, M.; Cumming, I. W.; Street, M. Sorption of Trace Heavy Metals by Thiol Containing Chelating Resins. *Solv. Extr. Ion Exch.* **2000,** *18* (1), 133–167.

17. Rawat, J. P.; Iqbal, M. Metal Ion Chelation Chromatography on Complex on Sorbed Stannic Silicate. *J. Liquid Chromatogr.* **1980,** *3* (11), 1657–1668.

18. Kim, H.-J.; Kim, J.-H.; Kim, W.-II; Suh, D. J. Nanoporous Phloroglucinol-Formaldehyde Carbon Aerogels for Electrochemical Use. *Korean J. Chem. Eng.* **2005,** *22* (5), 740–744.

19. Jirglova, H.; Perez Cadenas, A. F.; Maldonado-Hadar, F. J. *Langmuir* **2009,** *25*, 2461–2466.

20. Socrates, G. *Infrared Group Frequencies*; Wiley: Hoboken, NJ, 1980; pp 40–45.

21. Sun, C.; Qu, R.; Ji, C.; Wang, Q.; Sun, Y.; Cheng, G. A Chelating Resin Containing S, N and O Atoms: Synthesis and Adsorption Properties for Hg(II). *Eur. Polym. J.* **2006,** *42* (1), 188–194.

22. Samal, S.; Acharya, S.; Dey, R. K.; Ray, A. R. Synthesis, Characterization and Metal Ion Uptake Studies of Chelating Resin Derived from Formaldehyde/ Furfuraldehyde Condensed Phenolic Schiff Base of 4,4'-Diaminophenylmethane and *o*-Hydroxyacetophenone. *J. Appl. Polym. Sci.* **2003,** *88* (2), 570–581.

23. Rivas, B. R.; Peric, I. M.; Villegas, S. Synthesis and Metal Ion Uptake Properties of Water-Insoluble Functional Copolymers: Removal of Metal Ions with Environmental Impact. *Polym. Bull.* **2010,** *65* (9), 917–928.

24. Prasad, H. H.; Popat, K. M.; Anand, P. S. Synthesis of Cross-linked Methacrylic Acid-Coethylene Glycoldimethacrylate Polymers for the Removal of Copper and Nickel from Water. *Indian J. Chem. Technol.* **2002,** *9*, 385–393.

25. Shah, B. A.; Shah, A. V.; Bhandari, B. N. Recovery of Transition Metal Ions from Binary Mixtures by Ion Exchange Column Chromatography Using Synthesized Chelating Ion Resin Derived from *m*-Cresol. *Asian J. Chem.* **2004,** *16*, 1801–1810.

26. Shah, B. A.; Shah, A. V.; Bhandari, B. N. Selective Elution Metal Ions on a New Chelating Ion Exchange Resin Derived from Substituted 8-Hydroxyquinoline. *Asian J. Chem.* **2003,** *15*, 117–125.

CHAPTER 12

CYCLOMETALATED LIGANDS AND THEIR Ir(III) COMPLEXES AS EFFICIENT LUMINESCENCE MATERIALS FOR ORGANIC LIGHT-EMITTING DIODES APPLICATION

MEHA J. PRAJAPATI and KIRAN R. SURATI*

Department of Chemistry, Sardar Patel University, Vallabh Vidyanagar 388120, Gujarat, India

Corresponding author. E-mail: kiransurati@yahoo.co.in

ABSTRACT

Organic light-emitting diodes (OLEDs) have gained great interest in the last years due to their potential for future flat panel display and solid state lighting applications. The OLED technology is now being commercialized as a multi–billion-dollar market. OLEDs are already used in small displays in cellular phones, car stereos, digital cameras, and so on. The rapidly growing market for OLED displays and lighting is driving research both in developing low-cost advanced materials and improving manufacturing processes. Many types of metal complexes are used for OLEDs fabrication as emissive materials. There are still many problems concerning the efficiency, stability, color saturation, and manufacturing cost of such materials. To reach high luminescence, metal complexes, and especially transition, metal complexes have been widely investigated. Here, we have mainly focused on the types of cyclometalated ligands used and we review specially the C^N type of cyclometalated iridium complexes that are used as phosphorescent dopants in OLEDs.

12.1 INTRODUCTION AND OVERVIEW

Organic light-emitting diodes (OLEDs) is a new technology but research and development (R&D) has been initiated in 1960s since the discovery of organic electroluminescence molecule emission of light under electrical field.[1-8] In 1987, the major breakthrough in R&D occurred when Tang and Vanslyke[4] were the first to demonstrate OLED. OLEDs have been used as displays for electronic devices, lighting, decoration, and portable displays such as mobile phone, tablets, smart watches, and so on.

The future holds tremendous opportunity for this technology because of its low cost and high performance. For fabrication of OLEDs, many types of materials are utilized such as hole transporting, electron transporting, hole blocking, electron blocking, and emissive materials. Among these, this chapter mainly focuses on emissive materials (phosphorescence-based iridium complexes). Here, we have discussed on cyclometalation reaction and various types of cyclometalated ligand used. Also, a brief review is presented on C^N type of cyclometalated iridium complexes that are used as phosphorescent dopants in OLEDs.

12.2 CYCLOMETALATION REACTION AND ITS SCOPE

The term "cyclometalation" was introduced by Trofimenko, which describes reactions of transition metal complexes in which an organic ligand undergoes intramolecular metalation and forms a metal-carbon σ-bond, which are called cyclometalation reactions (Fig. 12.1). In 1970, Parshall focused on the activation of C-H bonds by transition metals.[9]

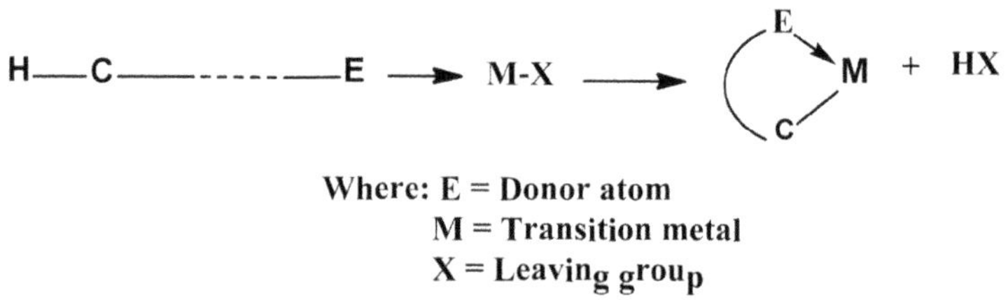

Where: E = Donor atom
M = Transition metal
X = Leaving group

FIGURE 12.1   General scheme for cyclometalation reaction.

12.2.1 SCOPE OF CYCLOMETALATION REACTIONS

Three main points are taken into consideration when forming the cyclo-metalated complexes, namely, (a) nature of metal atom, where the transition elements present in group VI to VIII of the periodic table forms metal–carbon σ–bonds with the exception of few. Mainly complexes containing the ruthenium, iridium, palladium, and so on, easily forms cyclometalated derivatives and are more commonly used; (b) nature of the metalated system; and (c) effect of donor atom.[10]

12.3 CLASSIFICATION OF CYCLOMETALATED LIGANDS

Generally, there are four types of cyclometalated ligands:
  I.   C^C
  II. C^N
  III. N^N
  IV. C^P

I. C^C
The first type of cyclometalated ligands consists of N-heterocyclic carbenes (NHC) such as pmb, pmi, and dfb-mb (Fig. 12.2), and are denoted as C^C ligands.

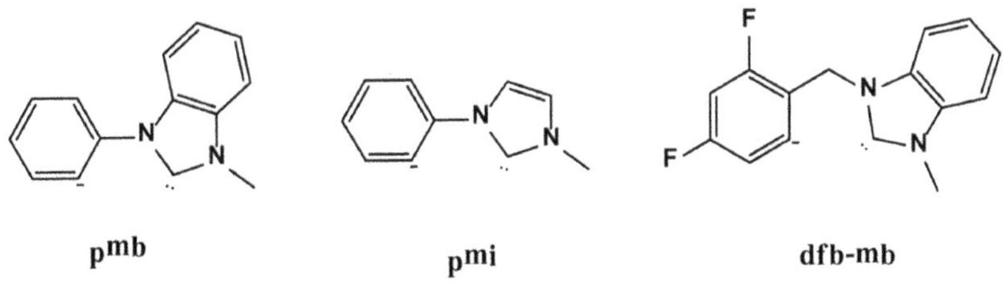

pmb   pmi   dfb-mb

FIGURE 12.2   Structures of pmb, pmi, and dfb-mb.

In principle, the carbene moiety stands for a neutral, two electron donor, which makes these cyclometalated ligands classical bidentate monoanionic ligands. Such type of C^C ligands exhibit a large ligand field strength with further destabilization of the d* energy states of the metal-centered system, which, in general, becomes inaccessible to emitting triplet, which reduces the quenching and hence can be used for designing various blue emitters. Here, it is also worth to note that the average distance between the metal–carbon (M–C) of the carbene is shorter than the respective metal–nitrogen (M–N) distances associated with the neutral nitrogen donor. This shorter M–C distance confirms that the carbene moiety affords much stronger ligand-to-metal dative interactions. The resulting larger ligand field strength would further destabilize the metal-centered d–d excited states, which could become thermally inaccessible from the typical emissive triplet excited state, making the carbene complexes suitable to serve as efficient phosphors.[10]

II. C^N

The second type of cyclometalated ligands (C^N) generally refers to the coordination of a heteroaromatic chelate to a metal element via a covalent metal–carbon bond and the remaining dative bonds are usually derived from heteroatoms such as nitrogen in the ligand. The most common example of C^N ligands is the 2-phenylpyridine (ppy; Fig. 12.3) that can bind to the variety of transition metal ions. As a result, five-membered metallacycle is formed through activation of the ortho C–H bond of the phenyl ring that is adjacent to the 2-pyridyl fragment.[10] All the cyclometalates are anionic as thistransformationinvolves removal of a proton fromthe aromatic heterocycles. Hence, this ligand offers very strong M–C covalent interactions as well as highly stabilized ligand-field strength toward the metal ion.

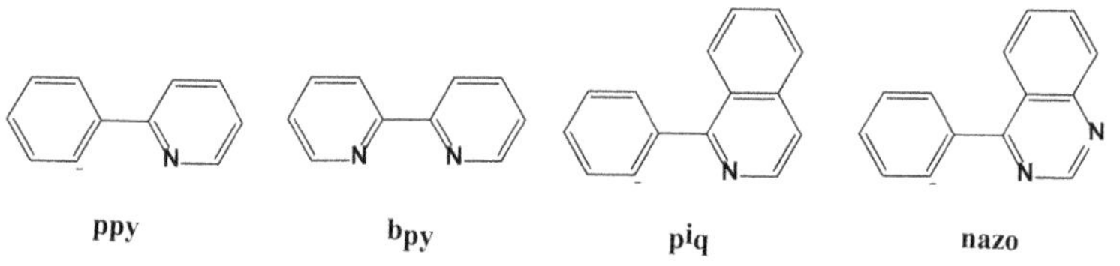

ppy   bpy   piq   nazo

FIGURE 12.3   Structures of ppy, bpy, piq, and nazo.

Furthermore, C^N ligand plays an important role in fine-tuning the emission wavelength of the metal complexes because of the π–π* energy gap. One can achieve red shift by reduction of the energy gap by the following means: (a) by incorporating electron releasing sulfur-containing heterocycles[11] (b) by extending the π-conjugation,[12,13] or (c) by adding of an electron accepting/releasing substituent or atom at the pyridyl (or phenyl) fragment.[14–16] Hence, C^N kind of ligand is important for the ππ* band gap modulation because the conjugation length can easily be increased or decreased, which will further change the band gap.

III. N^N

The third type of cyclometalated ligand consists of 2-pyridyl-C-linked azoles and are denoted as N^N chelates (Fig. 12.4). They are capable of using two of their nitrogen atoms to implement the required M–C bonding interaction.[17] Two highly electronegative nitrogen atoms of such N^N chelates enlarges the ligand-centered π–π* energy gap when comparing the phenyl or 4,6-difluorophenyl segments to the azole fragment. Hence, this enlarged π–π* energy gap provides many practical advantages in assembling luminescent transition-metal complexes. Hence, we can say that the N^N type cyclometalated ligands have a band gap from the π–π* orbitals (LC system) higher than the C^N type. The typical examples of pyridyl azolate chelates are listed as follows:

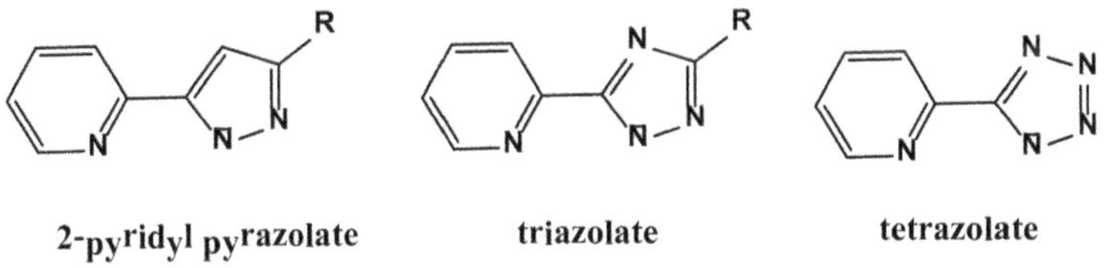

2-pyridyl pyrazolate   triazolate   tetrazolate

FIGURE 12.4   Structures of 2-pyridyl-C-linked azoles.

IV. C^P

The fourth type of cyclometalated ligand consists of PPh$_2$ (Fig. 12.5) group to increase metal-to-ligand charge transfer (MLCT) phosphorescence energy.

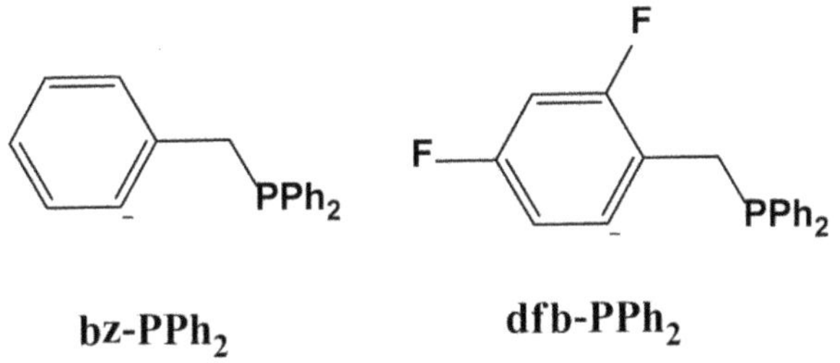

bz-PPh$_2$ dfb-PPh$_2$

FIGURE 12.5   Structures of bz-PPh$_2$ and dfb-PPh$_2$.

Also, they can lead to relatively strong ligand field strength with some further possibility of destabilization of the d* energy states.[18] The phosphino fragment can also undergo pre-coordination to the metal and then induce subsequent C–H activation of adjacent aromatic C–H bonds under mild conditions. For benzyldiphenylphosphine derivatives, selective formation of five-membered metallacycles occurs at the benzyl site, which provides a much more stable five-membered metal–chelate bonding interaction and, consequently, an easy access to the emissive metal complexes.[10] On the other hand, in case of triphenylphosphine derivates, the subsequent C–H activation results in the formation of highly unstable four-membered metallacycles due to the bond strain created, which is undesirable for constructing luminescent complexes.[10]

12.4 REVIEW OF SUCH COMPLEXES USED FOR OLEDS

Out of the four types of cyclometalated ligands discussed earlier, here we mainly focus on the complexes of iridium based on C^N type of cyclometalated ligands. In this section, we discussed many reviews and the research work related to Ir(III) complexes having various emissions like blue, green, orange, yellow, and red.

Kang et al. designed and synthesized novel iridium complex (bt)$_2$Ir(btz) (Fig. 12.6) by using 2-phenylbenzo[d]thiazole (bt) as cyclometalated ligand and 2-(2-hydroxyphenyl)benzothiazolato (btz) as ancillary ligand, which gives orange–yellow phosphorescence and measured its photophysical and electroluminescence properties for OLEDs applications. They used btz as ancillary ligand to tune the emission properties and the optical properties revealed that the π-conjuation in btz ligand is responsible for a red shift in emission. In addition, (bt)$_2$Ir(btz) is suitable for the low-cost solution processed production of orange–yellow phosphors for OLEDs.[19]

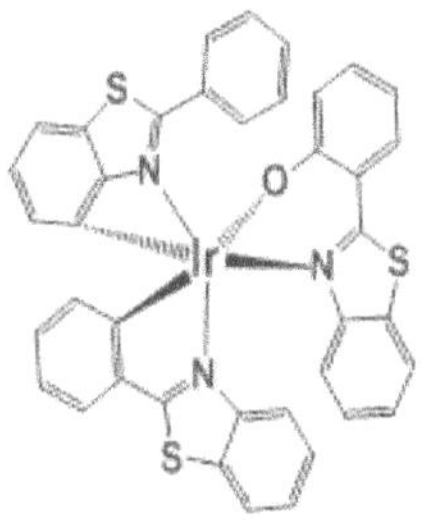

(bt)₂Ir(btz)

FIGURE 12.6 Structure of (bt)₂Ir(btz).

Han et al. synthesized two iridium complexes, namely, Ir(tfmpiq)₂(tpip) and Ir(tfmpzq)₂(tpip) (Fig. 12.7) using isoquinoline and quinazoline derivatives as cyclometalated ligands and tetraphenylimidodiphosphinate (tpip) as ancillary ligand and the result suggested that the iridium complex with quinazoline units are more potential materials for OLEDs as orange–red phosphors with high quantum yield. The device based on Ir(tfmpzq)₂(tpip) displayed better performance with maximum luminance of 129,466 cd/m², maximum current efficiency of 62.96 cd/A, and a maximum power efficiency of 53.43 lm/W with low efficiency roll-off.[20]

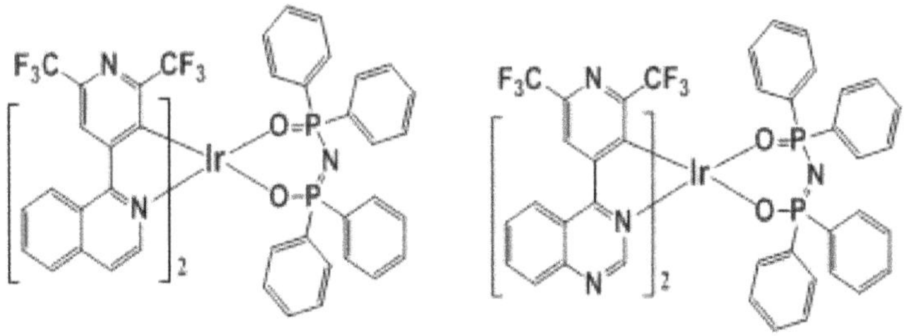

Ir(tfmpiq)₂(tpip)   Ir(tfmpqz)₂(tpip)

FIGURE 12.7 Structures of Ir(tfmpiq)₂(tpip) and Ir(tfmpzq)₂(tpip).

Zhang et al. designed and synthesized three iridium complexes, namely, [Ir(bzq)₂(POXD)], [Ir(pq)₂(POXD)], and [Ir(piq)₂(POXD)] (Fig. 12.8) using 7,8-benzoquinoline (bzq), 2-phenylquinoline (pq), and

1-phenylisoquinoline (piq) as cyclometalated ligand and N-(5-phenyl-1,3-4-oxadiazol-2-yl)-diphenylphosphinic amide (POXD) as ancillary ligands. Complexes give emission in the range of 539–614 nm, that is, from yellow to red light having quantum yields from 0.06 to 0.21. The study reveals that the complex having bzq as cyclometalated ligand shows excellent performance with maximum current efficiency of 70.1 cd/A, maximum external quantum efficiency of 21.3%, and maximum luminance of 24080 cd/m² respectively. Also, POXD as ancillary ligand would have potential applications in highly efficient phosphorescent OLEDs.[21]

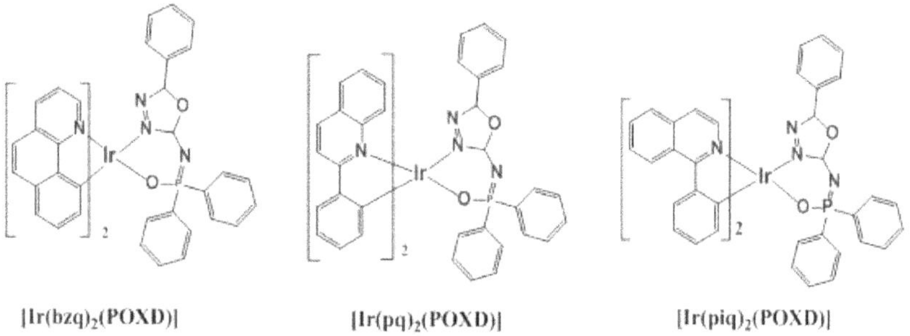

|Ir(bzq)₂(POXD)| |Ir(pq)₂(POXD)| |Ir(piq)₂(POXD)|

FIGURE 12.8   Structures of [Ir(bzq)₂(POXD)], [Ir(pq)₂(POXD)], and [Ir(piq)₂(POXD)].

A homoleptic Ir(III) complex fac-Ir(SFXpy)₃ was synthesized by Ren et al. in 2017 using spiroligand containing spiro[fluorene-9,9'-xanthene] unit (Fig. 12.9). fac-Ir(SFXpy)₃ gives yellow emission at 542 nm and exhibits good solubility in common organic solvents, which is favorable for solution-based device fabrication for OLEDs. The complex shows good thermal stability having decomposition temperature of 356°C. The result from wet-processed device shows current efficiency of 32.2 cd/A, power efficiency of 22.1 lm/W, and external quantum efficiency of 11.3%, respectively.[22]

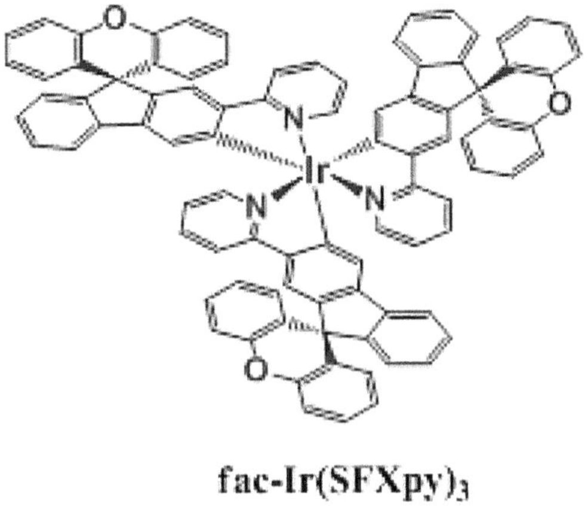

fac-Ir(SFXpy)₃

FIGURE 12.9   Structure of fac-Ir(SFXpy)₃.

Ivanov et al. synthesized iridium complex (Cl-bt)₂Ir(acac) (Fig. 12.10) using 2-(4-chlorophenyl)benzothiazole as cyclometalated ligand and acetlyacetone as the ancillary ligand, which is used as a dopant in the hole transporting layer of an OLEDs. Depending on the dopant concentration fine tuning of color from blue-greenish to orange is achieved. The complex gives emission at 543 nm, which is blue shifted as compared to (bt)₂Ir(acac) which gives emission at 558 nm.[23]

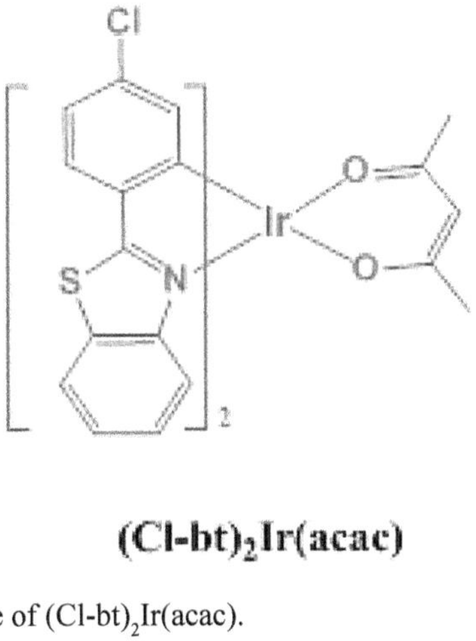

(Cl-bt)₂Ir(acac)

FIGURE 12.10   Structure of (Cl-bt)₂Ir(acac).

Wang et al. synthesized homoleptic Ir(III) complexes containing 2-phenylbenzothiazole derivatives as ligands (Fig. 12.11). For the first time, they introduced various electron-deficient and electron-rich groups, like $-CH_3$, $-OCH_3$, and fluorine onto the 6-position of benzothiazole moiety and studied their optical and electrochemical properties. The result reveals that introducing electron-rich groups in the ligand makes oxidation process shift slightly to more positive potential for these complexes, while reverse is observed for electron-releasing groups. All the complexes show emission in the range of 540–546 nm having quantum yield ranging from 29% to 37%.[24]

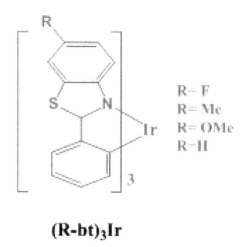

R– F
R= Me
R= OMe
R–H

(R-bt)₃Ir

FIGURE 12.11   Structures of (R-bt)₃Ir.

Orange phosphorescent iridium complexes were designed and synthesized by Cho et al. using 1-(8-naphthyl)-isoquinoline as the cyclometalated ligand and 2-phenylpyridine (ppy) and 2-phenylquinoline (pq) as ancillary ligands. The complexes (8-NAIQ)Ir(ppy)₂ and (8-NAIQ)Ir(pq)₂ (Fig. 12.12) shows yellow to orange emission at 536 nm and 583 nm, respectively and are desirable as orange emitters for two emitting component white organic light emitting diodes (WOLEDs).[25]

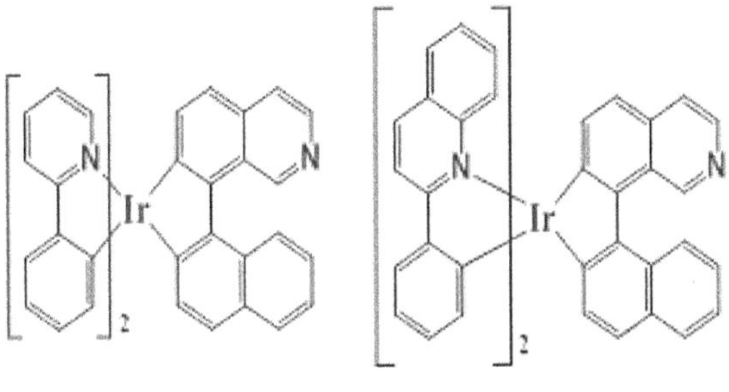

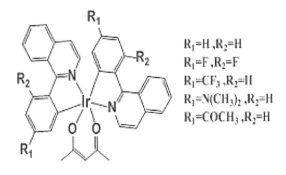

(8-NAIQ)Ir(ppy)₂   (8-NAIQ)Ir(pq)₂

FIGURE 12.12 Structures of (8-NAIQ)Ir(ppy)$_2$ and (8-NAIQ)Ir(pq)$_2$.

A series of novel iridium complexes of 1-phenylisoquinoline deriva-tives (Fig. 12.13) were synthesized and characterized by F. Ali et al. Optical properties of these complexes reveals that they gave emission in the orange to the deep red region having λ_{PL} ranging from 598 to 658 nm. The quantum yield and the lifetime of these complexes were found in the range of 0.1 to 0.32 μs and 0.43 to 1.9 μs, respectively. The device result shows the maximum brightness of 7600 cd/m^2 and current efficiency of ~7.0 cd/A indicating that these complexes are promising materials for OLEDs.[26]

R_1=H ,R_2=H
R_1=F ,R_2=F
R_1=CF$_3$,R_2=H
R_1=N(CH$_3$)$_2$,R_2=H
R_1=COCH$_3$,R_2=H

Ir(R)(piq)₂

FIGURE 12.13 Structure of Ir(R)(piq)$_2$.

Dumur et al. synthesized novel heteroleptic Ir(III) complex Ir(piq)$_2$(dbm) using 1-phenylisoquinoline (piq) as cyclometalated ligand and dibenzoylmethane (dbm) as ancillary ligands which can be used as an emitter in phosphorescent OLEDs (Fig. 12.14). The photoluminescent spectra of the complex were measured in dichloromethane solution, which gives red emission at 623 nm having quantum yield of 0.069 and luminescence lifetime of 130 ns. The device performance was tested with different dopant concentration were the highest brightness was obtained with 8 wt % dopant and a luminance of 7475 cd/m^2, current efficiency of 6.2 cd/A.[27]

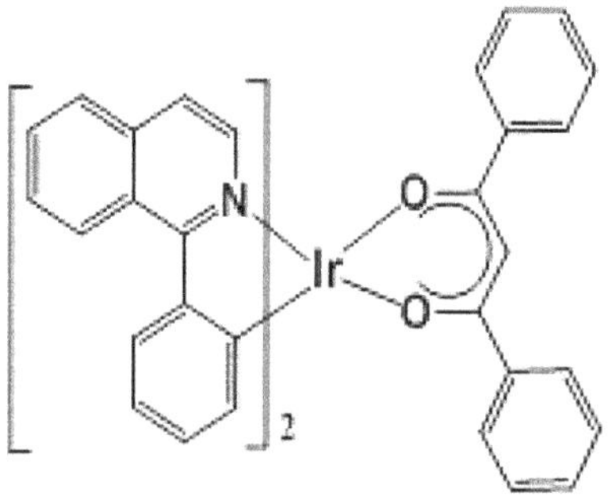

Ir(piq)$_2$(dbm)

FIGURE 12.14 Structure of Ir(piq)$_2$(dbm).

By using fluorinated 1-phenylisoquinoline derivatives as the cyclometalated ligands and acetlyacetone as ancillary ligands (Fig. 12.15), Park et al. synthesized efficient red emitters for OLEDs. Density functional theory (DFT) study was performed to calculate the energy levels of HOMO and LUMO of these iridium complexes. These complexes show PL emission bands in the range of 599–608 nm having bandgap in the range of 2.88–3.13 Ev, respectively. They observed that rather than electron withdrawing nature of fluorine in 1-phenylisoquinoline, the substituent position of fluorine had a significant effect on the emission wavelength and the energy bandgap.[28]

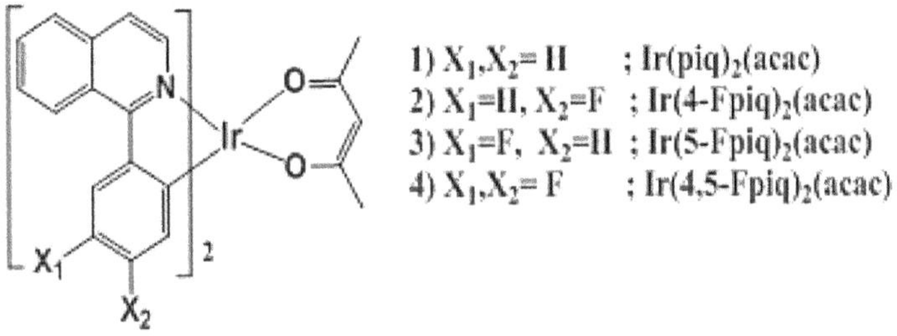

1) $X_1, X_2 = H$; Ir(piq)$_2$(acac)
2) $X_1 = H, X_2 = F$; Ir(4-Fpiq)$_2$(acac)
3) $X_1 = F, X_2 = H$; Ir(5-Fpiq)$_2$(acac)
4) $X_1, X_2 = F$; Ir(4,5-Fpiq)$_2$(acac)

Ir(piq)$_2$(X)

FIGURE 12.15 Structures of Ir(piq)$_2$(X).

Okada and co-workers synthesized Ir(III) complexes by using 1-phenyl-isoquinoline as the main ligand and studied the effect of substituent of these complexes (Fig. 12.16). The result reveals that the emission wavelength and quantum yield of these complexes differs depending on the substituents. These complexes show red emission ranging from 598 to 635 nm, and quantum yield and lifetime in the ranges of 17–32% and 0.97–2.34 µs, respectively. They also found that the complexes having electron withdrawing groups have higher oxidation potential when compared with the electron donating groups. The complex Ir4F5Mpiq shows excellent power efficiency of 12.4 lm/W and an external quantum efficiency of 15.5 %.[12]

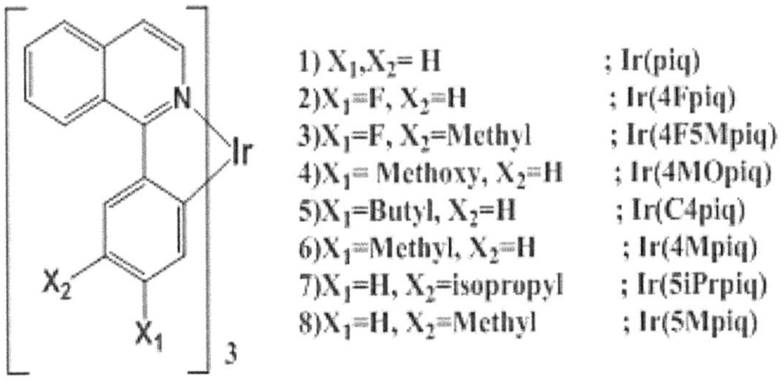

1) $X_1, X_2 = H$; Ir(piq)
2) $X_1 = F, X_2 = H$; Ir(4Fpiq)
3) $X_1 = F, X_2 = Methyl$; Ir(4F5Mpiq)
4) $X_1 = Methoxy, X_2 = H$; Ir(4MOpiq)
5) $X_1 = Butyl, X_2 = H$; Ir(C4piq)
6) $X_1 = Methyl, X_2 = H$; Ir(4Mpiq)
7) $X_1 = H, X_2 = isopropyl$; Ir(5iPrpiq)
8) $X_1 = H, X_2 = Methyl$; Ir(5Mpiq)

Ir(X-piq)$_2$

FIGURE 12.16 Structures of Ir(X-piq)$_2$.

Li and Zhou et al. synthesized efficient green OLEDs based on 2-(4-(tris-fluoromethyl)phenylpyridine as the cyclometalated ligand and the tetra-(4-trifluoromethylphenyl)imidodiphosphinate) as the ancillary ligands (Fig. 12.17). They showed that the introduction of (–CF$_3$) moiety to the iridium complexes shows good green emission both in solid and solution state having lifetime of 1.88 µs. The result of the devices confirms that the complex can be used as green phosphorescent emitter that has potential application in OLEDs.[29]

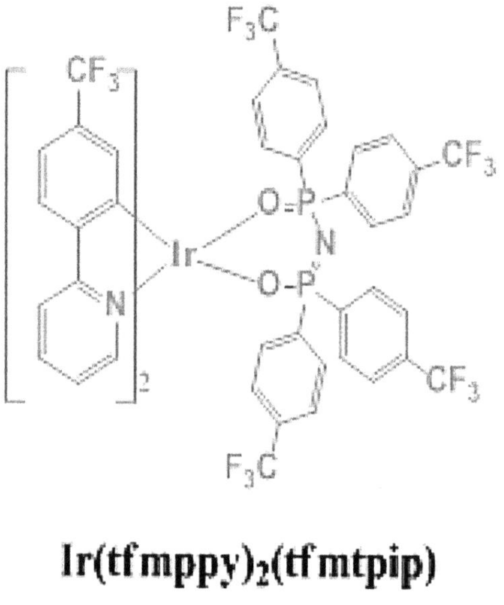

Ir(tfmppy)$_2$(tfmtpip)

FIGURE 12.17   Structure of Ir(tfmppy)$_2$(tfmtpip).

Sudhir and co-workers reported three novel heteroleptic Ir(III) complexes using 2-phenylpyridine as the cyclometalated ligand and 5-hydroxy-3-methyl-1-phenyl-1H-pyrazole-4-carbaldehyde as the ancillary ligands (Fig. 12.18). These complexes shows green emission in the range of 511–512 nm, where as the complex Ir(ppy)$_2$L1 shows highest quantum yield of 89% having excitons lifetime of 0.34 µs. All the three complexes show good thermal stability.[30]

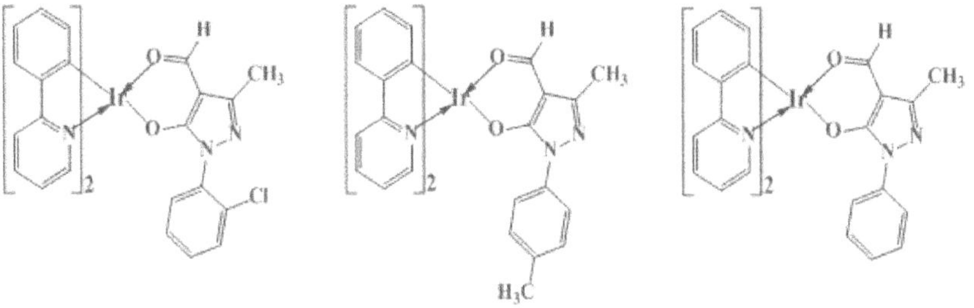

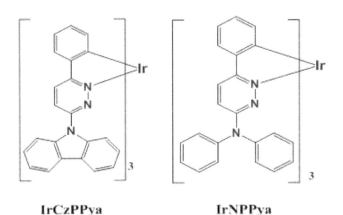

Ir(ppy)₂L1   Ir(ppy)₂L2   Ir(ppy)₂L3

FIGURE 12.18 Structures of Ir(ppy)₂L1, Ir(ppy)₂L2, and Ir(ppy)₂L3.

By using carbazole and diphenylamine as the substituents, Guo et al. designed and synthesized two homoleptic complexes, IrCzPPya and IrNPPya, which shows strong orange emission (Fig. 12.19). Both the complexes exhibit good thermal stability and gives orange emission having λ_{PL} and Φ_{PL} in the range of 556–558 nm and 0.45–0.55, respectively. The device achieved maximum efficiency of 55.9 lm/W, 49.9 cd/A, and 23.9% with CIE coordinates of (0.33, 046), which are among the best with similar CIE coordinates.[31]

IrCzPPya   IrNPPya

FIGURE 12.19 Structures of IrCzPPya and IrNPPya.

Aravind and Shivakumar designed and synthesized new heteroleptic Ir(III) complexes: Ben-Ir-ppy, Ben-Ir-acac, and Ben-Ir-pic by using

2-phenylpyridine (ppy), acetlyacetone (acac), and 2-picolinic acid (pic) as ancillary ligand (Fig. 12.20) and investigated its photophysical and electrochemical properties. In addition, all these complexes emerging out as promising luminescent materials and its hole transporting performance is more favorable than electron transport performance. The observed quantum yields for these iridium complexes were found in the range of 0.12–0.29 in solution. Also, according to theoretical calculations, it is found that, complex having acetlyacetone as the ancillary ligand is a more potential emitter for OLED applications.[32]

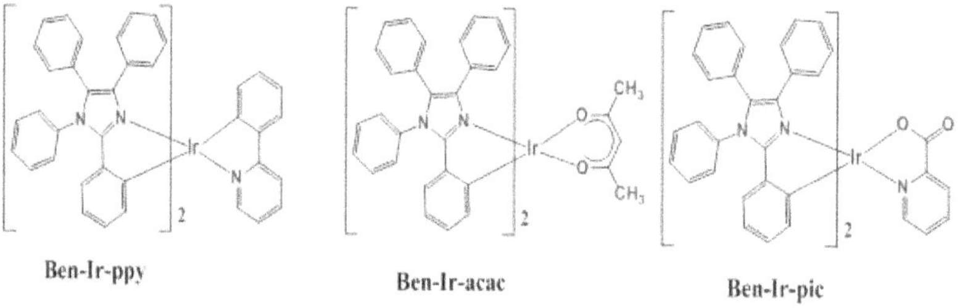

Ben-Ir-ppy   Ben-Ir-acac   Ben-Ir-pic

FIGURE 12.20   Structures of Ben-Ir-ppy, Ben-Ir-acac, and Ben-Ir-pic.

12.5   CONCLUSION

- In this article, first a brief introduction of OLED is discussed.
- We have discussed about the cyclometalation reaction and types of cyclometalated ligands used in literature
- We reviewed the research work related to Ir(III) complexes having various emissions like blue, green, orange, yellow, and red specially based on C^N type of cyclometalated ligands.

In summary, such ligands mainly contribute to tune the photophysical as we as electrochemical properties of complexes. Besides this, it also offers the following:

- The usual ligand strong field with metal ions leads to an increase of the band gap from dd* orbitals to a further reduced probability of having populated high energy d* states. As discussed previously, such population is responsible for the typical phosphorescence quenching by promoting of the nonradiative decay.

- The existence of LC transition in a near-energy location will help the color tuning proses

ACKNOWLEDGMENTS

Authors express sincere thanks to Science and Engineering Research Board (SERB), New Delhi SB/EMEQ-204/2014 and Gujarat Council of Science and Technology (GUJCOST), Gandhinagar (Ref. No. GUJCOST/MRP/14-15/27). Scheme for Young Scientist and Technologist (SYST) (Ref.: SP/YO/008/2017(G)); New Delhi, Meha J. Prajapati acknowledges financial support as INSPIRE fellowship (DST/INSPIRE Fellowship/2015/IF 150296) from the Department of Science and Technology (DST), New Delhi.

KEYWORDS

- **cyclometalation**
- **cyclometalated ligands**
- **iridium complexes**
- **organic light-emitting diodes**
- **luminescence**

REFERENCES

1. Pope, M.; Kallmann, H. P.; Magnante, P. *J. Chem. Phys.* **1963**, *38*, 2042.
2. Helfrichj, W.; Schneider, W. G. *Phy. Rev. Lett.* **1964**, *14*, 229.
3. Miyata, S. *Organic Electroluminescent Materials and Device;* 1997. ISBN: 9782919875108.
4. Tang, C. W.; VanSlyke, S. A. *Appl. Phys. Lett.* **1987**, *51*, 913.
5. Pode, R.; Diouf, B. *Solar Lighting;* Springer-Verlag: London, 2011.
6. Kayne, R.; Sell, H. *Revolution in Lamps: A Chronicle of. 50 Years of Progress*, The Fairmont Press, 2nd Edition,Georgia, **2001**.
7. Bernanose, A.; Vouaux, P. *J. Chim. Phys.* **1953**, *52*, 509.
8. VanSlyke, S. A.; Tang, C. W. Organic Electroluminescent Devices Having Improved Power Conversion Efficiencies." US Patent: 4539507, 1985.
9. Purahull, G. W. *Acc. Chem. Res.* **1975**, *8*, 113.

10. Chi, Y.; Chou, P. T. *Chem. Soc. Rev.* **2010,** *39*, 638.

11. Lamansky, S.; Djurovich, P.; Murphy, D.; Abdel-Razzaq, F.; Lee, H. -E.; Adach, C.; Burrows, P. E.; Forrest, S. R.; Thompson, M. E. *J. Am. Chem. Soc.* **2001,** *123*, 4304.

12. Okada, S.; Okinaka, K.; Iwawaki, H.; Furugori, M.; Hashimoto, M.; Mukaide, T.; Kamatani, J.; Igawa, S.; Tsuboyama, A.; Takiguchi, T.; Ueno, K. *Dalton Trans.* **2005,** *9,* 1583.

13. Mi, B. X.; Wang, P. F.; Gao, Z. Q.; Lee, C. S.; Lee, S. T.; Hong, H. L.; Chen, X. M.; Wong, M. S.; Xia, P. F.; Cheah, K. W.; Chen, C. H.; Huang, W. *Adv. Mater.* **2009,** *21*, 339.

14. Hwang, F. M.; Chen, H. Y.; Chen, P. S.; Liu, C. S.; Chi, Y.; Shu, C. F.; Wu, F. I.; Chou, P. T.; Peng, S. M.; Lee, G. H. *Inorg. Chem.* **2005,** *44*, 1344.

15. Zhou, G.; Ho, C. L.; Wong, W. Y.; Wang, Q.; Ma, D.; Wang, L.; Lin, Z.; Marder, T. B.; Beeby, A. *Adv. Funct. Mater.* **2008,** *18*, 499.

16. Zhou, G.; Wang, Q.; Ho, C. L.; Wong, W. Y.; Ma, D.; Wang, L.; Lin, Z. *J. Chem. Asian* **2008,** *3*, 1830.

17. Cheng, C. C.; Yu, W. S.; Chou, P. T.; Peng, S. M.; Lee, G. H.; Wu, P. C.; Song, Y. H.; Chi, Y. *Chem. Commun.* **2003,** *20,* 2628.

18. Pereira, L. *Organic Light Emitting Diodes: The Use of Rare-Earth and Transition Metals;* CRC Press, 2012. ISBN: 978-9-81426-795-3.

19. Kang, S. K.; Jeon, J.; Jin, S. H.; Kim, Y. I. *Bull. Korean Chem. Soc.* **2017,** *38*, 646.

20. Han, H. B.; Cui, R. Z.; Jing, Y. M.; Lu, G. Z.; Zheng, Y. X.; Zhou, L.; Zuo, J. L.; Zhang, H. *J. Mater. Chem. C* **2017,** *1*, DOI: 10.1039/C7TC02117H.

21. Zhang, F.; Si, C.; Dong, X.; Wei, D.; Yang, X.; Guo, K.; Wei, B.; Li, Z.; Zhang, C.; Li, S.; Zhai, B.; Cao, G.; *J. Mater. Chem. C* **2017**. DOI: 10.1039/C7TC02420G.

22. Ren, B. Y.; Guo, R. D.; Zhong, D. K.; Ou, C. J.; Xiong, G.; Zhao, X. H.; Sun, Y. G.; Jurow, M.; Kang, J.; Zhao, Y.; Li, S. B.; You, L. X.; Wang, L. W.; Liu, Y.; Huang, W. *Inorg. Chem.* **2017**. DOI: 10.1021/acs.inorgchem.7b01034.

23. Ivanov, P.; Tomova, R.; Petrova, P.; Stanimirov, S.; Petkov, I. *J. Phys.* **2014,** *514*, 012038. DOI: 10.1088/1742-6596.

24. Wang, R.; Liu, D.; Ren, H.; Zhang, T.; Wang, X.; Li, J. *J. Mater. Chem.* **2011,** *21*, 15494.

25. Cho , S. H.; Ha, Y. *Mo. Crystals and Liquid Crystals* **2017,** *644*, 234.

26. Ali, F.; Nayak, P. K.; Periasamy, N.; Agarwal, N. *J. Chem. Sci.* **2017,** *129* (9), 1391.

27. Dumur, F.; Lepeltier, M.; Siboni, H. Z.; Xiao, P.; Graft, B.; Lalevee, J.; Gigmes, D.; Aziz, H. *Thin Solid Films* **2014**. DOI: 10.1016/j.tsf.2014.04.096.

28. Park, G. Y.; Seo, J. H.; Kim, Y. K.; Kim, Y. S.; Ha, Y. *Japanese J. Appl. Phys.* **2007,** *46* (4B), 2735.

29. Li, H. Y.; Zhou, L.; Teng, M. T.; Xu, Q. L.; Lin, C.; Zheng, Y. X.; Zuo, J. L.; Zhang, H. J.; You, X. Z. *J. Mater. Chem. C* **2013,** *1*, 560.

30. Kumar, S.; Surati, K. R.; Lawrence, R.; Vamja, A. C.; Yakunin, S.; Kovalenko, M. V.; Santos, E. J. G.; Shih, C. J. *Inorg. Chem.* **2017,** *56* (24), 15304.

31. Guo, L. Y.; Zhang, X. L.; Wang, H. S.; Liu, C.; Li, Z. G.; Liao, Z. J.; Mi, B. X.; Zhou, X. H.; Zheng, C.; Lia, Y. H.; Gao, Z. Q. *J. Mater. Chem. C* **2015**. DOI: 10.1039/c5tc00458f.

32. Kajjam, A. B.; Sivakumar, V. *J. Photochem. Photobiol. A*. DOI: 10.1016/j.jphotochem.2017.09.054.

CHAPTER 13

FACILE SYNTHESIS OF SOME TRIAZINE-BASED CHALCONES AS POTENTIAL ANTIOXIDANT AND ANTIDIABETIC AGENTS

RAVINDRA S. SHNIDE[*]

Department of Chemistry and Industrial Chemistry, Dayanand Science College, Latur 413512, Maharashtra, India

[*]*E-mail: rshinde.33381@gmail.com*

ABSTRACT

A series of s-triazine-based chalcones have been prepared by the Claisen–Schmidt condensation. Chalcones have characteristic 1,3-diaryl-2-propen-1-one backbone skeleton. Changes in their aryl rings have access of a high degree of variety that has proven useful for the development of new medicinal agents with improved potency and lesser toxicity. A convenient method for the synthesis of biologically active triazine-based chalcones using triazine ketone and substituted benzaldehyde in dry methanol has been done. The structures of the compounds were confirmed by spectral data (Infra Red (IR), Proton Magnetic Resonance (^{1}H NMR), and mass spectroscopy). The synthesized compounds were studied for their antioxidant and antidiabetic activity.

13.1 INTRODUCTION

Chalcones are α,β-unsaturated ketones with the reactive keto ethylenic group –CO–CH=CH–. The aromatic chalcones are recognized as phenyl styryl ketones and they are colored because of the occurrence of the chromophore –CO–CH=CH–. Their color depends on the presence of other auxochromes in the structure. The first name "chalcone" was given by Kostanecki and Tambor. The chalcones are very good quality synthons because of this reason a variety of new heterocycles with good pharmaceutical property can be designed. Thurston et al. revealed that the triazine nucleus containing chalcones and their derivatives have their own importance in heterocyclic chemistry.[1–3] They are extremely useful intermediates for the synthesis of five-, six-, and seven-membered heterocyclic compounds. The chalcone is very reactive with various reagents such as hydrazine hydrate, phenyl hydrazine, 2-amino thiophenol, and so on.

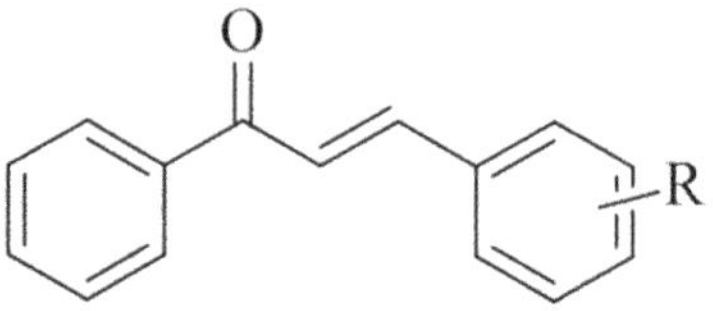

Aromatic Chalcones

The variety of chalcones are used to manufacture a large number of derivatives like pyrazolines, cyanopyridines, isoxazoles, pyrimidines, flavonoids, isoflavonoids, aurones, tetralones, aziridines, clavicins, and so on. The vital role of chalcones includes their use in purification of blood, monitoring cholesterol level, regulating blood pressure, strengthen the immune system, preventing thrombus and cancer, promoting metabolism, and suppressing acid secretion. It is well known that most natural or synthetic chalcones are highly active with pharmaceutical and medicinal application.[2,3]

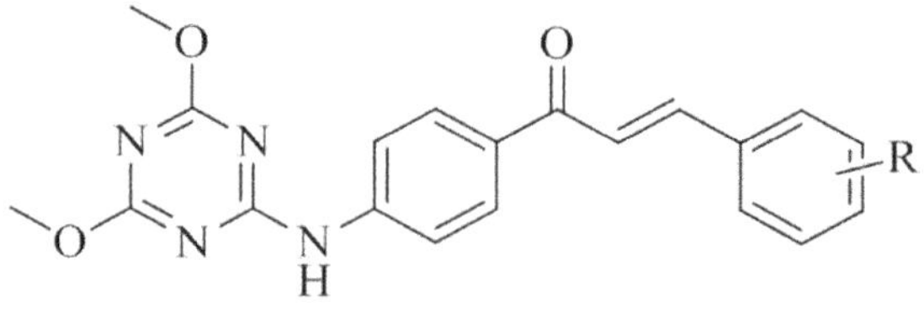

Scaffold of Triazine Chalcone

Lee et al. and his coworker shows the convenience of manipulating the triazine scaffold and the diverse biological properties are useful to design a fusion library by combining a fluorophore and biophore utilizing chalcone and triazines.[4-6] They found to be active as antimicrobial,[7] antibacterial,[8] antimalerial,[9-10] antileishmanial,[11] antiprotozoal,[12] antitubercular,[13] anticancer,[14-15] antifungal,[16] anti-inflammatory,[17] and good antibacterial activity.[18] The chemistry of chalcones has been recognized as a significant field of study. Chalcones have been studied extensively because of their wide range of biological activity.[19-25] Various methods are available for the synthesis of chalcones. One of the methods for synthesis of 1,3-diaryl-2-propenones (chalcones) involves the use of strong bases like potassium hydroxide, sodium hydroxide, $Ba(OH)_2$, LiHMDS, and phosphates. Some other methods are reported to make use of are acid catalysed aldol condensations involving acids like $AlCl_3$, dry HCl, $ZrCl_2/NiCl_2$, and $RuCl_3$ (for cyclic and acyclic ketones). The most suitable method is the Claisen–Schimdt condensation of the equimolar quantities of aryl ketone with aryl aldehyde in the presence of alcoholic alkali.

Considering the significance of triazine-based chalcones and vast scope for further studies, we herein report the synthesis of aryl substituted   1-[(4-(4,6-dimethoxy-1,3,5-triazin-2-yl)aminophenyl)]3-phenylprop-2-en-1-one. The synthesis involves the reaction between 2-chloro-4,6-dimethoxy-1,3,5-triazine and 4-aminoacetophenone in dry acetone to form an intermediate 1-(4-(4,6-dimethoxy-1,3,5-triazin-2-yl) aminophenyl) ethanone which on further treatment with various aromatic aldehydes in presence of 40% NaOH in dry methanol solvent at room temperature gives the required triazine chalcone (Scheme 13.1). Structures of the synthesized compounds were ascertained by IR , NMR, and mass-spectral analysis. All resultant compounds were screened for their antioxidant and antidiabetic activity.

13.2   PRESENT WORK

The present scheme involves the synthesis of 1-(4-(4,6-dimethoxy-1,3,5-triazin-2-yl) aminophenyl)-3-phenylprop-2-en-1-one by the three-step process as illustrated in Scheme 13.1.

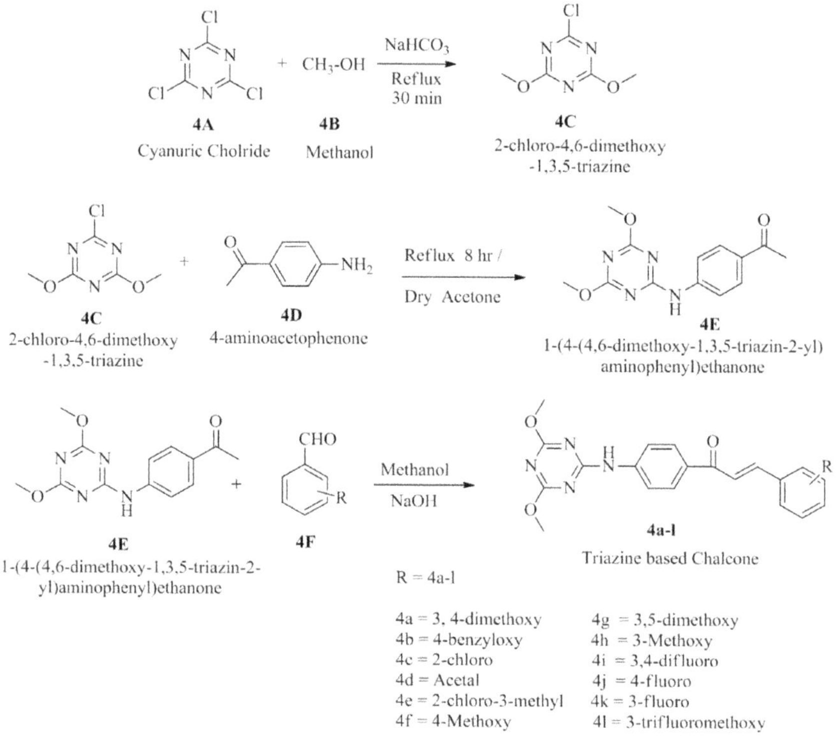

SCHEME 13.1 Synthesis of aryl substituted 1-[(4-(4,6-dimethoxy-1,3,5-triazin-2-yl) minophenyl)]-3-phenylprop-2-en-1-one (**4a–l**).

The first step comprises synthesis of 2-chloro-4,6-dimethoxy-1,3,5-tri-azine (**4c**) by the reaction of cyanuric chloride (**4a**) with methanol (**4b**) in the presence of $NaHCO_3$ reflux for 30 min. The resulting intermediate (**4c**) is formed in very good yield (70–75%) by the nucleophilic displacement of two chlorine atoms of cyanuric chloride by methoxy anion. In the second step, the synthesis of amine derivative of intermediate, that is, compound **4e** was achieved with 74% of yield by reaction between intermediate **4c** and 4-aminoacetophenone (**4d**) in dry acetone at 80–90°C for 8 h. In the third step, the subsequent coupling of the compound (**4e**) with the various aromatic aldehydes (**4f**) under a basic condition involving 40% NaOH in dry methanol solvent at room temperature yielded the corresponding aryl substituted 1-[(4-(4,6-dimethoxy-1,3,5-triazin-2-yl)aminophenyl)]3-phenylprop-2-en-1-one (**4a–l**) as final compound. The reaction proceeded in good yield.

In addition, the biological effect of the presence or absence of various functional groups on aromatic aldehyde is also studied. The antiantioxidant activity study of synthesized compounds was undertaken to investigate the biological potential of triazine chalcones. It is done by 2,2-diphenyl-1-picrylhydrazyl (DPPH·) radical scavenging method. The antidiabetic activity of the compounds was investigated by estimating the degree of nonenzymatic hemoglobin glycosylation method.

13.2.1 MECHANISM FOR SYNTHESIS OF ARYL-1-[(4-(4,6-DIMETHOXY-1,3,5-TRIAZIN-2-YL) AMINO PHENYL)]-3-PHENYLPROP-2-EN-1-ONE (4A–L)

13.3 EXPERIMENTAL

The cyanuric chloride, 4-aminoacetophenone, and substituted benzaldehyde were purchased from Sigma Aldrich Chemicals Pvt. Ltd., Mumbai, India. The Thin-layer chromatography (TLC) plates (silica gel 60F254) were obtained from Merck, Germany. All melting points were resoluted by open-capillary method and are uncorrected. The IR spectra were recorded with a nexus 470FT-IR spectrophotometer at Department of Chemistry, Dayanand College, Solapur. The ^{1}H NMR spectra were recorded in DMSO-d_6 on a Bruker DRX-400 MHz and the mass spectra were obtained on Jeol-SX-102 (FAB) spectrometer at SIAF Indian Institute of Technology, Bombay.

13.3.1 SYNTHESIS OF 1-[(4-(4,6-DIMETHOXY-1,3,5-TRIAZIN-2-YL)AMINOPHENYL)]3-PHENYLPROP-2-EN-1-ONE (4A–L)

The target compounds were prepared in the following three steps:

Step I: Synthesis of 2-chloro-4,6-dimethoxy-1,3,5-triazine (4c):

In the first step, cyanuric chloride (4a) (18.5 g, 0.1 mol) was dissolved in 80 mL methanol (4b) and 5 mL water in a 250-mL round bottom flask and sodium bicarbonate (16.8 g, 0.2 mol) was added slowly in the reaction mixture at room temperature. Then, the reaction mixture was refluxed for 30 min until the evolution of CO_2 was stopped. The contents were poured onto ice-cold water and filtered. The white shiny solid product (4c) as an intermediate was obtained and recrystalized from dichloromethane, dried in desiccators (Scheme 13.1).

Yield: 13 g (74%)   MP: 74–76°C

^{1}**H NMR** (400 MHz, DMSO-d_6): δ 4.05 (s, 6H, 2-OCH$_3$)

Mass Spectroscopy (MS) Electrospray ionization (ESI) m/z: 175.67 [M + 1]$^+$, 177.67 [M + 2]$^+$

Anal. data: Calcd. for $C_5H_6ClN_3O_2$: C, 34.20; H, 3.44; N, 23.93%; found: C, 33.98; H, 3.34; N, 23.70%.

Step II: Synthesis of 1-(4-(4,6-Dimethoxy-1,3,5-triazin-2-yl) aminophenyl) ethanone (4e)

4-Aminoacetophenone (**4d**) (1.35 g, 0.01 mol) and 2-chloro-4,6-dime-thoxy-1,3,5-triazine (**4c**) (1.75 g, 0.01 mol) were dissolved in dry acetone (50 mL). The reaction mixture was refluxed for 8 h. The sodium carbonate solution (0.005 N, 10 mL water) was added to neutralized HCl generated during condensation. After completion of reaction, which is confirmed on thin-layer chromatography, the reaction mixture was poured into ice water. The solid separated was filtered, washed with water, dried, and purified by ethyl acetate and hexane to get pure product (**4e**), that is, 1-[(4-(4,6-dimethoxy-1,3,5-triazin-2-yl)aminophenyl]ethanone.

Physicochemical Data

Yield: 76%   MP: 205–207°C.

IR (KBr cm⁻¹): 3302 (NH– stretch), 1766 (–C=O), 808 (–C–N= str. *s*-triazine)
   ¹H NMR (400 MHz, DMSO-d_6) δ 3.92 (s, 6*H*, –2OCH$_3$), 2.45 (s, 3*H*, –COCH$_3$), 7.80–8.05 (m, 4*H*, Ar–H), 10.45 (s, 1*H*, Ar–NH)
   MS (ESI) *m/z*: 275.28 [M + 1]⁺; Anal. data: Calcd for C$_{13}$H$_{14}$N$_4$O$_3$: C, 56.93; H, 5.14; N, 20.43%; found: C, 56.80; H, 5.08; N, 20.17%

Step III: Synthesis of 1-[4-(4,6-Dimethoxy-1,3,5-triazin-2-yl) aminophenyl)]-3-(3,4-dimethoxyphenyl)prop-2-en-1-one (4a)

The compound (**4e**) (2.74 g, 0.01 mol) was dissolved in methanol (40 mL) and 3,4-dimethoxybenzaldehyde (1.66 g, 0.01 mol) (**4f**) was added with constant stirring at room temperature for 30 min, then sodium hydroxide (40% w/v) was added to reaction mixture which was again stirred at room temperature for 16 h. The progress of the reaction was monitored by TLC (20% hexane:ethyl acetate). After completion of the reaction, crushed ice was added in the reaction mixture and neutralized with HCl. The product separated was filtered, washed with water, dried, and recrystalized from ethanol to get pure product (**4a**).

Similarly, compounds (**4b–l**) were prepared following the same procedure.

13.4 RESULTS AND DISCUSSION

The 2-chloro-4,6-dimethoxy-1,3,5-triazine was prepared by reaction of cyanuric chloride (**4a**) (0.1 mol) with sodium bicarbonate (0.2 mol) in methanol (**4b**) and water. Further, compound (**4c**) react with 4-amino acetophenone (**4d**) gives the compound (**4e**). Subsequent condensation of triazine ketone (**4e**) with various aldehydes (**4f**) led to compound (**4a–l**). The structures of synthesized compounds were assigned on the basis of spectral characterization IR, ^{1}H NMR, mass spectral, and elemental analysis.

13.4.1 CHARACTERISTIC TEST FOR 1-[(4-(4,6-DIMETHOXY-1,3,5-TRIAZIN-2-YL)AMINOPHENYL)]3-PHENYLPROP-2-EN-1-ONE (TRIAZINE CHALCONES)

All triazine chalcones gave a positive test with Wilson reagent and red coloration with conc. H_2SO_4.

13.4.1.1 WILSON TEST

Wilson reagent was freshly prepared by mixing the saturated acetonic solution of boric acid and 10% acetonic solution of citric acid. About 0.001 g of triazine chalcones was dissolved in 1 mL acetone and to that 1–2 mL Wilson reagent was added. The chalcone gave strong yellow coloration indicating a positive test for the presence of chalcone moiety.

13.4.1.2 IR SPECTRA

IR spectra of triazine chalcones (**4a–l**) showed distinguishing band at close to region 1620–1650 cm^{-1} due to >C=O stretching vibration. A decrease of standard unsaturated >C=O band to lesser wave number is recognized due to the occurrence of α,β-unsaturated double bond. All chalcones showed characteristics absorption band in the region 1500–1650 cm^{-1} due to olefinic (–CH=CH–) group.

13.4.1.3 ^{1}H NMR SPECTRA

The ^{1}H NMR of compound (**4a–l**) showed doublet for –CO–CH=CH– at 6.90–7.10 ppm and doublet for Ar—CH=CH– at 7.80–8.30 ppm which confirm the presence of ethylenic double bond, respectively. The olefinic α,β-unsaturated proton shift at the downfield region is the characteristic of the system.

13.4.1.4   MASS SPECTRA

The mass spectra for aryl substituted 1-[(4-(4,6-dimethoxy-1,3,5-triazin-2-yl)aminophenyl)]3-phenylprop-2-en-1-one   (**4a–l**)   showed   intense molecular ion peaks followed by cleavage on both sides of the carbonyl group.

The analytical, IR, and ^{1}H NMR data for compounds (**4a–l**) are given in Table 13.1–13.3.

TABLE 13.1   Analytical and Mass Spectral Data of 1-[(4-(4,6-Dimethoxy-1,3,5-triazin-2-yl)aminophenyl)]3-phenylprop-2-en-1-one **4(a–l)**.

Sr. no.	Elemental analysis found (calculated)			Mass (m/z)	MP (°C)	Yield (%)
	% C	% H	% N			
4a	62.38 (62.55)	5.08 (5.25)	13.13 (13.26)	423.32 [M + 1]$^+$	110–112	75
4b	62.08 (62.22)	5.11 (5.16)	11.83 (11.96)	469.26 [M + 1]$^+$	116–118	75
4c	60.44 (60.53)	4.15 (4.32)	14.03 (14.12)	397.21 [M + 1]$^+$, 399.2 [M + 2]	140–142	73
4d	61.98 (62.05)	4.20 (4.46)	13.53 (13.79)	407.31 [M + 1]$^+$	218–220	72
4e	61.20 (61.39)	4.29 (4.66)	13.43 (13.64)	411[M + 1]$^+$ 412.9 [M + 2]	136–138	72
4f	64.05 (64.28)	5.09 (5.14)	14.03 (14.28)	393 [M + 1]$^+$	122–124	75
4g	62.44 (62.55)	5.15 (5.25)	13.13 (13.26)	423.17[M + 1]$^+$	112–114	75
4h	64.28 (64.15)	5.04 (5.14)	14.13 (14.28)	393.16 [M + 1]$^+$	123–124	74
4i	60.15 (60.30)	4.00 (4.05)	13.98 (14.05)	399.13 [M + 1]$^+$	138–140	72
4j	63.05 (63.15)	4.34 (4.50)	14.50 (14.73)	381.10 [M + 1]$^+$	129–131	74
4k	63.05 (63.15)	4.26 (4.50)	14.59 (14.73)	381.13 [M + 1]$^+$	130–132	74
4l	56.44 (56.50)	3.56 (3.84)	12.65 (12.77)	447.14 [M + 1]$^+$	150–152	74

TABLE 13.2 IR Spectral Data of Aryl Substituted 1-[(4-(4,6-dimethoxy-1,3,5-triazin-2-yl)aminophenyl)]3-phenylprop-2-en-1-one (**4a–l**).

Sr. no.	IR band position (wave number cm^{-1})					
	–NH str. (s) in 2° amine	–C–H str. (s) in –OCH$_3$	–CH=CH– str. (m)	>C=O str. (s) in chalcone	C–O–C str.(m) ether	–C–N– str. in triazine
4a	3278	2945	1590	1650	1237	803 (s), 1332 (w)
4b	3335	2942	1592	1662	1250	811, 1330
4c	3299	2952	1628	1670	1260	810, 1338
4d	3330	2846	1641	1672	1250	810, 1340
4e	3350	2850	1633	1670	1239	806, 1337
4f	3350	2987	1640	1665	1220	807, 1336
4g	3370	2976	1646	1678	1250	810, 1335
4h	3366	2950	1649	1672	1226	806, 1334
4i	3275	2961	1647	1671	1235	812, 1340
4j	3369	2963	1649	1674	1233	801, 1338
4k	3410	2960	1630	1678	1259	809, 1336
4l	3396	2969	1636	1682	1257	805, 1338

S, strong, *m*, medium, and *w*, weak.

TABLE 13.3 ^{1}H NMR Data for Aryl Substituted 1-[(4-(4,6-dimethoxy-1,3,5-triazin-2-yl) aminophenyl)]3-phenylprop-2-en-1-one (**4a–l**).

Sr. no.	Assignments, number of protons, multiplicity, chemical shift in ppm, and coupling constant *J* in Hz				
	–OCH$_3$ in triazine	–CO–CH=CH–/ Ar–CH=CH– in chalcones	Ar–OCH$_3$/– OCH$_2$–	Ar–H in aromatic	–NHAr
4a	6*H*, s (3.99)	1*H*, d (7.96) *J* = 12.8 Hz, 1*H*, d (8.23) *J* = 12.8 Hz	3*H*, s (3.95), 3*H*, s (3.97)	2*H*, d (7.02) *J* = 8 Hz, 2*H*, dd (7.38–7.40) *J* = 8 Hz, 1*H*, s (7.54), 1*H*, d (7.78–7.76) *J* = 8.1 Hz, 1*H*, d (7.82) *J* = 8.1 Hz	1*H*, s (10.45)
4b	6*H*, s (3.88)	1*H*, d (7.96) *J* = 12.7 Hz, 1*H*, d (8.18) *J* = 12.7 Hz	2*H*, s (5.18)	2*H*, d (7.10), 9*H*, m (7.21–7.50), 2*H*, d (7.85)	1*H*, s (10.12)

TABLE 13.3 *(Continued)*

Sr. no.	Assignments, number of protons, multiplicity, chemical shift in ppm, and coupling constant J in Hz				
	–OCH$_3$ in triazine	–CO–CH=CH–/ Ar–CH=CH– in chalcones	Ar–OCH$_3$/– OCH$_2$–	Ar–H in aromatic	–NHAr
4c	6H, s (3.98)	1H, d (7.98) J = 12.8 Hz, 1H, d (8.20) J = 12.8 Hz	–	8H, m (7.40–7.60)	1H, s (10.50)
4d	6H, s (3.98)	1H, d (7.95) J = 12.8 Hz, 1H, d (8.16) J = 12.8 Hz	2H, s (6.10)	2H, dd (7.00) J = 8.1 Hz, 2H, dd (7.32) J = 8.1 Hz, 2H, m (7.65–7.80), 1H, s (7.88)	1H, s (10.40)
4e	6H, s (3.98) 3H, s (2.40) Ar–CH$_3$	1H, d (8.08) J = 12.8 Hz, 1H, d (8.19) J = 12.8 Hz	–	3H, m (7.32– 7.48), 4H, m (7.90–8.00)	1H, s (10.41)
4f	6H, s (3.99)	1H, d (7.98) J = 12.7 Hz, 1H, d (8.18) J = 12.7 Hz	3H, s (3.82)	8H, m (7.10–7.90)	1H, s (10.47)
4g	6H, s (3.96)	1H, d (7.92) J = 12.8 Hz, 1H, d (8.15) J = 12.8 Hz	3H, s (3.94), 3H, s (3.96)	2H, d (7.10) J = 8 Hz, 2H, dd (7.35–7.41) J = 8 Hz, 1H, s (7.52) 1H, d (7.76–7.72) J = 8.1 Hz, 1H, d (7.83) J = 8.1 Hz	1H, s (10.41)
4h	6H, s (3.96)	1H, d (7.88) J = 12.8 Hz, 1H, d (8.16) J = 12.8 Hz	3H, s (3.84)	8H, m (7.10–7.90)	1H, s (10.46)
4i	6H, s (3.93)	1H, d (7.93) J = 12.8 Hz, 1H, d (8.21) J = 12.8 Hz	–	7H, m (7.10–7.82)	1H, s (10.22)
4j	6H, s (3.96)	1H, d (7.91) J = 12.8 Hz, 1H, d (8.22) J = 12.8 Hz	–	8H, m (7.12–7.89)	1H, s (10.38)
4k	6H, s (3.94)	1H, d (7.89) J = 12.8 Hz, 1H, d (8.19) J = 12.8 Hz	–	8H, m (7.15–7.95)	1H, s (10.35)
4l	6H, s (3.98)	1H, d (7.91) J = 12.8 Hz, 1H, d (8.23) J = 12.8 Hz	–	8H, m (7.20–7.98)	1H, s (10.38)

13.5 BIOLOGICAL EVALUATION

The antioxidant and antidiabetic activity was done as per literature.[26-31]

13.5.1 ANTIOXIDANT ASSAY

The antioxidant effect of aryl substituted 1-[(4-(4,6-dimethoxy-1,3,5-tri-azin-2-yl)aminophenyl)]3-phenylprop-2-en-1-one (**4a–l**) was estimated by free radical scavenging. To determine the free radical scavenging activity, a 0.1-mM solution of DPPH radical in methanol was prepared and 1 mL of this solution was added to 3 mL of the test material (**4a–l**) at different concentrations prepared in methanol by Huang et al.[29] The solutions were incubated for 30 min at room temperature and then absorbance was measured at 517 nm. The decreasing of the DPPH solution absorbance indicates an increase in the DPPH radical-scavenging activity. This activity is given as % DPPH radical scavenging that is calculated in the equation using DPPH solution as a control. The ascorbic acid compound was used as a standard drug for assay. The observed data for screening are presented in Table 13.4.

% DPPH radical scavenging = 1 − [CA − SA/CA] × 100

where CA is the control absorbance and SA is the sample absorbance.

13.5.2 ANTIDIABETIC ASSAY (NONENZYMATIC GLYCOSYLATION OF HEMOGLOBIN ASSAY)

The antidiabetic activities of aryl substituted 1-[(4-(4,6-dimethoxy-1,3,5-triazin-2-yl)aminophenyl)]3-phenylprop-2-en-1-one (**4a–l**) were investigated by estimating degree of nonenzymatic hemoglobin glycosylation, which is measured[30] colorimetrically at 520 nm. Glucose (2%), hemoglobin (0.06%), and sodium azide (0.02%) solutions were prepared in phosphate buffer 0.01 M, pH 7.4. One milliliter each of the above solution was mixed. One milliliter of each test solution (**4a–l**) (final concentration 10 µg/mL) of the sample was added to the above mixture. The mixture was incubated in dark at room temperature for 72 h. The degree of glycosylation of hemoglobin was measured colorimetrically at 520 nm. Alpha-tocopherol (Trolax) was used as a standard drug for assay. The percentage of inhibition was calculated as per the previously published protocol. All the tests were performed in triplicate. The observed data for screening are presented in Table 13.4.

TABLE 13.4 Antioxidant and Antidiabetic Activity of Aryl Substituted 1-[(4-(4,6-dimethoxy-1,3,5-triazin-2-yl)aminophenyl)]3-phenylprop-2-en-1-one (**4a–l**).

Compounds sr. no.	Antioxidant activity	Antidiabetic activity
	% DPPH radical scavenging	% inhibition
4a	29.36	12.5
4b	49.12	19.3
4c	56.4	15.12
4d	55.21	23.85
4e	23.22	29.27
4f	27.5	10.23
4g	29.25	12.3
4h	27.4	10.20
4i	30.35	28.50
4j	32.22	30.55
4k	38.55	32.56
4l	45.30	33.22
Ascorbic acid	72.25	NA
α-Tocopherol	NA	46.28

NA, not applicable.

13.5.3 STRUCTURE–ACTIVITY RELATIONSHIP OF ANTIOXIDANT ACTIVITY

The equilibrium between the creation and removal of free radicals determines the general stability of a living body. Free-radical chain reactions in the body are initiated frequently by reactive species (RS– ions, free radicals, reactive nitrogen species, molecules, reactive oxygen species) possessing oxygen or nitrogen atom with an unpaired electron. The free radical-scavenging activities of aryl substituted 1-[(4-(4,6-dimethoxy-1,3,5-triazin-2-yl)aminophenyl)]3-phenylprop-2-en-1-one (**4a–l**) was evaluated by DPPH scavenging method. All the synthesized compounds (**4a–l**) exhibit potential free radical-scavenging ability. Among these synthesized compounds, some compounds (**4b**, **4c**, **4d**, and **4l**) shows comparable activity with standards antioxidant, ascorbic acid which ranges from 45.3% to 56.4%. Remaining compounds have shown moderate activity as compared with the standard antioxidant, which ranges from 23.22% to 38.5%. From the results in Table 13.4, it was observed that the presence of strong electronic withdrawing substituent plays a very important role in antioxidant activity while the presence of strong electronic donating substituent is not so favorable for antioxidant property. From the above results, it was concluded that compounds possessing electronic withdrawing groups $-OCF_3$, acetal, benzyloxy, and chloro at *ortho/para/meta* position of phenyl ring derived from benzaldehyde showed promising antioxidant activity. The results obtained in the present study are in line with other findings. Thus in a broader sense, the presence of lipophilic groups such as chloro, $-OCF_3$, at 3 and 5 positions have favorable effects on the antioxidant activity. The spatial, lipophilic, and structural properties of the compounds determined their antioxidant properties. Due to the presence of the reactive ketovinylenic group, chalcones, and their analogs have been reported to be antioxidant and hydroxyl and phenyl substituents are associated with antioxidant property.[24]

13.5.4 STRUCTURE–ACTIVITY RELATIONSHIP OF ANTIDIABETIC ACTIVITY

The aryl-substituted 1-(4-(4,6-dimethoxy-1,3,5-triazin-2-yl)aminophenyl) 3-phenylprop-2-en-1-one (**4a–l**) with chlorine, methyl, methoxy, and fluoro-substitutions at position 2, 3, 4, 5 on phenyl ring derived from benzaldehyde exhibited the moderate antidiabetic activity as compared to α-tocopherol. The structure–activity relationship (SAR) of the tested chalcones was studied and the findings were supported by the use of antidiabetic clinical drug, α-tocopherol as an optimistic control. The percentage of inhibition was calculated as previously published protocol. The results are shown in Table 13.4. SAR analysis data are shown to represent different substitutions on phenyl ring derived from benzaldehyde of the tested chalcones and represents the effect of substitution on culture medium glucose concentration. Triazine-based chalcones with $-OCF_3$, $-F$, $-Cl$, and $-CH_3$ substitutions at position 3, 3, 4, 3, 4, and 2, 3, respectively, on phenyl ring derived from benzaldehyde **4e**, **4i**, **4j**, **4k**, and **4l** exhibited good activity with culture glucose medium concentrations. Additionally, triazine chalcones with methoxy substitution at position 3, 4, and 5 on phenyl ring derived from benzaldehyde (**4a**, **4f**, **4g**, and **4h**) show lower activity. It is noteworthy to state that benzyloxy, chloro, and acetal substitution on phenyl ring of benzaldehyde affected chalcones activity **4b**, **4c**, and **4d** to a moderate extent. The results showed that substitution of chloro, fluoro, difluoro, trifluoro, and methoxy at position 2, 3, 4 of phenyl ring of benzaldehyde, significantly affected the glucose uptake activity. These data suggest that the glucose uptake activity of chalcone is significantly affected by fluoro substitution at position 2, 3, 4 on phenyl ring derived from benzaldehyde. From these results, it is observed that triazine chalcones **4e**, **4i**, **4j**, **4k**, and **4l** showed good antidiabetic activity while other triazine chalcone **4a**, **4b**, **4c**, **4d**, **4f**, **4g**, and **4h** showed moderate to lower antidiabetic activity. The results showed that the substitution of fluoro, chloro, bromo, and hydroxy at position 2 of the ring, significantly affected the glucose uptake activity in the cell model.[25]

13.6   CONCLUSION

In this research work, we have successfully synthesized a series of compounds comprising of aryl substituted 1-[(4-(4,6-dimethoxy-1,3,5-triazin-2-yl)aminophenyl)]3-phenylprop-2-en-1-one by Claisen–Schmidt condensation of the equimolar quantities of aryl ketone with aryl aldehyde in presence of methanolic alkali. This is the most suitable method for synthesis of triazine chalcones at room temperature. The process offers an easy reaction profile and simple workup procedure. The synthetic yields of the generated products ranged from 72% to 75% and their structures were established by spectral technique (IR, NMR, and MS).

Finally, all of the synthesized compounds have been tested for their antioxidant and antidiabetic activities. The compounds **4b**, **4c**, **4d**, and **4l** triazine chalcones exhibited comparable antioxidant activity with standard compound ascorbic acid (72.25%) which ranged from 45.3% to 56.4% and remaining showed a moderate activity of 23.22–38.5%. This was found due to electronic withdrawing groups –OCF$_3$, acetal, benzyloxy, and fluoro at *ortho/para/meta* position of phenyl rings present in these compounds.

The compounds exhibited comparable antidiabetic activity (10.23–33.22%) with standard compound α-tocopherol (46.25%). The triazine-based chalcones with $-OCF_3$, $-F$, $-Cl$, and $-CH_3$ substitutions at position 3, 3, 4, 3, 4, and 2, 3, respectively, on phenyl ring derived from benzaldehyde **4e**, **4i**, **4j**, **4k**, and **4l** exhibited good activity with culture glucose-medium concentrations. Additionally, triazine chalcones with methoxy substitution at position 3, 4, 5 on phenyl ring derived from benzaldehyde (**4a**, **4f**, **4g**, and **4h**) show lower activity. It is noteworthy to state that benzyloxy, chloro, and acetal substitution on ring of benzaldehyde affected chalcones activity **4b**, **4c**, and **4d** to a moderate extent. The results showed that substitution of chloro, fluoro, difluoro, trifluoro, and methoxy at position 2, 3, 4 of the phenyl ring of benzaldehyde, significantly affected the glucose uptake activity. These data suggest that the glucose uptake activity of chalcone is significantly affected by fluoro-substitution at position 2, 3, 4 on phenyl ring derived from benzaldehyde. From these results, it is observed that triazine chalcones **4e**, **4i**, **4j**, **4k**, and **4l** showed good antidiabetic activity while other triazine chalcone **4a**, **4b**, **4c**, **4d**, **4f**, **4g**, and **4h** showed moderate to lower antidiabetic activity.

KEYWORDS

- antioxidant
- antidiabetic activity
- cyanuric chloride
- 2-chloro-4,6-dimethoxy-1,3,5-triazine
- substituted benzaldehyde
- triazine chalcone

REFERENCES

1. Thurston, J. T.; Dudley, J. R.; Kaiser, D. W.; Hechenbleikner, I.; Schaefer, F. C.; Holm-Hansen, D. Cyanuric Chloride Derivatives I. Amino-*s*-triazines. *J. Am. Chem. Soc.* **1951,** *73*, 2981–2983.
2. Kaiser, D. W.; Thurston, J. T.; Dudley, J. R.; Schaefer, F. C.; Hechenbleikner, I.; Holm-Hansen, D. Cyanuric Chloride Derivatives. II. Substituted Melamines. *J. Am. Chem. Soc.* **1951,** *73*, 2984–2986.

3. Dudley, J. R.; Thurston, J. T.; Schaefer, F. C.; Holm-Hansen, D.; Hull, C. J.; Adams, P. Cyanuric Chloride Derivatives III. Alkoxy-*s*-triazines. *J. Am. Chem. Soc.* **1951**, *73*, 2986–2990.

4. Wang, S.; Lee, W. S.; Ha, H. H.; Chang, Y. T. Combinatorial Synthesis of Galactosyl-1,3,5-Triazines as Novel Nucleoside Analogues. *Org. Biomol. Chem.* **2011**, *9*, 6924–6926.

5. Khersonsky, S. M.; Chang, Y. T. Safety-Catch Approach to Orthogonal Synthesis of a Triazine Library. *Comb. Chem.* **2004**, *6*, 474–477.

6. Hsieh, H. K.; Tsao, L. T.; Lin, C. N. Synthesis and Anti-inflammatory Effect of Chalcones. *J. Pharm. Pharmacol.* **2000**, *52*, 163–171.

7. Viana, G. S.; Bandeira, M. A.; Matos, F. J. Analgesic and Antiinflammatory Effects of Chalcones Isolated from *Myracrodruon urundeuva* Allemão. *Phytomedicine* **2003**, *10*, 189–195.

8. Zhao, L. M.; Jin, H. S.; Sun, L. P.; Piao, H. R.; Quan, Z. S. Synthesis and Evaluation of Antiplatelet Activity of Trihydroxychalcone Derivatives. *Bioorg. Med. Chem. Lett.* **2005**, *15*, 5027–5029.

9. Lee, S.; Zhai, D.; Chang, Y. T. Development of a Chalcone Triazine Fusion Library: Combination of a Fluorophore and Biophore. *Tetrahedr Lett.* **2013**, *54*, 2976–2979.

10. Daukshas, V. K.; Ramamauskas, Y.; Udrenaite, A. B.; Brukshtus, V. V.; Lapinskas, R. S.; Maskalyunas, M. K. Synthesis and Local-Anesthetic Activity of 6-[ω-Amino-ω)-arylalkyl] Benzo-1,4-dioanes. *Pharm. Chem. J.* **1984**, *18*, 471–475.

11. Bremner, P. D.; Meyer, J. J. Pinocembrin Chalcone: An Antibacterial Compound from *Helichrysum trilineatum*. *Planta Med.* **1998**, *64* (8), 777.

12. Nielsen, S. F.; Boesen, M.; Larsen, k.; Schonning, H. K. Antibacterial Chalcones Bioisosteric Replacement of the 4′-Hydroxy Group. *Bioorg. Med. Chem.* **2004**, *12* (11), 3047–3054.

13. Kenyon, R.; Li, G. L.; Cohen, E. F.; Chen, X.; Gong, B.; Dominguez, J. N.; Davidson, K. G.; Miller, R. E.; Nuzum, E. O.; McKerrowis, J. H. In Vitro Antimalarial Activity of Chalcones and Their Derivatives. *J. Med. Chem.* **1995**, *38* (26), 5031–5037.

14. Liu, M.; Wilairar, P.; Go, M. L. Antimalarial Alkoxylated and Hydroxylated Chalcones: Structure–Activity Relationship Analysis. *J. Med. Chem.* **2001**, *44* (25), 4443–4452.

15. Go, M. L.; Liu, M.; Nilairat, P. J.; Rosental, K. J.; Saliba, K. J.; Kirk, K. Antiplasmodial Chalcones Inhibit Sorbitol-Induced Hemolysis of *Plasmodium falciparum*-Infected Erythrocytes. *Antimicrob. Agents Chemother.* **2004**, *48* (9), 3241–3245.

16. Zhai, L.; Chen, M.; Blom, T. G.; Theander, S, B.; Christensen, A.; Khazarmi, J. The Antileishmanial Activity of Novel Oxygenated Chalcones and Their Mechanism of Action. *Antimicrob. Chemother.* **1999**, *43*, 793–803.

17. Lunardi, F.; Guzela, A. T.; Rodrigues, R.; Eger-Mangrich, Correa, M.; Steindel, E. C.; Grizard, J.; Assreuy, J. B.; Calixto, A. R.; Santos, S. Trypanocidal and Leishmanicidal Properties of Substitution-Containing Chalcones. *Antimicrob. Agents Chemother.* **2003**, *47* (4), 1449–1451.

18. Lin, Y. M.; Zhou, Y.; Flavin, M. T.; Zhou, L. M.; Nie, W.; Chen, F. C. Chalcones and Flavonoids as Anti-Tuberculosis Agents. *Bioorg. Med. Chem.* **2002,** *10* (8), 2795–2802.

19. Modzelewska, A.; Pettit, C.; Achanta, G.; Davidson, N. E.; Huang, P.; Khan, S. R. Anticancer Activities of Novel Chalcone and Bis-chalcone Derivatives. *Bioorg. Med. Chem.* **2006,** *14* (10), 3491–3495.

20. Rao, Y. K.; Fang, S. H.; Tzeng, Y. M. Differential Effects of Synthesized 2′-Oxygenated Chalcone Derivatives: Modulation of Human Cell Cycle Phase Distribution. *Bioorg. Med. Chem.* **2004,** *12* (10), 2679–2686.

21. Svetaz, L.; Tapia, A.; Lopez, S.; Furlan, R. L. E.; Petenatti, E.; Pioli, R.; Schmeda, G. H.; Zacchino, S. A. Antifungal Chalcones and New Caffeic Acid Esters from *Zuccagnia punctata* Acting against Soybean Infecting Fungi. *J. Agric. Food Chem.* **2004,** *52* (11), 3297–3300.

22. Nowakowska, Z. A review of Anti-Infective and Anti-Inflammatory Chalcones. *Eur. J. Med. Chem.* **2007,** *42* (2), 125–137.

23. Certin, A.; Cansiz, A.; Digrak, M. 3-Aryl-5-furylpyrazolines and Their Biological Activities. *Heteroatom. Chem.* **2003,** *194*, 345–347.

24. Forkmann, G.; Heller, W. *Comprehensive Natural Products Chemistry*; Elservier Science: Amsterdam, 1999; p 713.

25. Villanova, P. A. *Approved Standard Document M-7A*; National Committee for Clinical Laboratory Standards (NCCLS), 1985.

26. Murray, P. R.; Baron, E.; Jorgensen, J.; Landry, M.; Pfaller, M. *Manual of Clinical Microbiology*, 9th ed., American Society of Microbiology: Washington, DC, 2007.

27. Kato, K.; Terao, S.; Shimamoto, N.; Hirata, M. Studies on Scavengers of Active Oxygen Species, Synthesis and Biological Activity of 2-*O*-Alkyl Ascorbic Acids. *J. Med. Chem.* **1998,** *31* (4), 793–798.

28. Padmaja, A.; Payani, T.; Reddy, G. D.; Padmavathi, V. Synthesis, Antimicrobial and Antioxidant Activities of Substituted Pyrazoles, Isoxazoles, Pyrimidine and Thioxopyrimidine Derivatives. *Eur. J. Med. Chem.* **2009,** *44* (11), 4557–4566.

29. Huang, D. J.; Lin, C. D.; Chen, H. J.; Lin, Y. H. Antioxidant and Antiproliferative Activities of Sweet Potato (*Ipomoea batatas*) Constituents. *Bot. Bull. Acad. Sin.* **2004,** *45*, 179–186.

30. Naderi, G. H.; Dinani, N. J.; Asgary, S.; Taher, M.; Nikkhoo, N.; Boshtam, M. Effect of Some High Consumption Spices on Hemoglobin Glycation. *Ind. J. Pharm. Sci.* **2014,** *76*, 553–557.

31. Ochoa, J. L. Pain Mechanisms in Neuropathy. *Curr. Opin. Neurol.* **1994,** *7*, 407–414.

CHAPTER 14

FUNDAMENTALS OF DISCHARGE PLASMAS

PRIJIL MATHEW[1,*], SAJITH MATHEWS T.[1], P. J. KURIAN[1], and AJITH JAMES JOSE[2]

[1]*Post Graduate and Research Department of Physics, St Berchmans College Changanassery, Kerala 686 101, India*

[2]*Post Graduate and Research Department of Chemistry, St Berchmans College Changanassery, Kerala 686 101, India*

Corresponding author. E-mail: prijilmk@gmail.com

ABSTRACT

Plasma is a typical complex medium of a collection of charged particles exhibiting a wide variety of linear and nonlinear phenomena. To understand the behavior of the plasma medium, a simple glow discharge system is chosen. Initial attempts are made to understand various mechanisms leading to gas breakdown. The term discharge means the flow of electric current through the gaseous medium. The easiest way to produce plasma is by applying voltage between two electrodes. To understand the behavior of any gas discharges, it is essential to know the detailed process involving with individual particles. This is so because even if sufficient electrons to carry the entire current are produced from some source, they cannot be expected to form the current without encountering the gas particles. Here, we describe the production of plasma by ionizing the gas in low pressure with the applied electric field. A systematic study of the fundamentals of glow discharge plasma was included in this chapter.

14.1   INTRODUCTION

Solids, liquids, and gases are the three commonly known states of matter. An extraordinarily unique state with unique properties called plasma is the fourth state of matter. Intensive studies on plasma have been carried out recently.[1] Dissociation (breaking of bonds) of a gas can be achieved by applying heat energy to the gas. The bond breaks to form positively charged ions and negatively charged electrons. This constitutes an ionized gas. The addition of more energy to the gas results in the ionization of more atoms.[2] At a temperature above 100,000 K, most matter exists in an ionized state; this ionized state of matter is called the fourth state or plasma state. A plasma state can exist at temperatures lower than 100,000 K, provided there is a mechanism for ionizing the gas, and if the density is low enough so that recombination is not rapid.[2]

The word "plasma" comes from Greek "πλασμα" which means something molded. To describe the behavior of ionized gas in an electrical discharge, Nobel laureate Irving Langmuir and Lewi Tonks in 1929 used the term "plasma."[2,3] Plasma is the fourth state of matter and the most natural state of matter (~99%) in the universe and is defined as "an ionized gas or a quasi-neutral gas in which the presence of unbound (freely moving) charged particles (usually electrons and positive ions) in large number along with neutral particles; each particle interacts simultaneously and collectively with its innumerable, immediate nearby, and distant neighbor charged particles through their electromagnetic fields."[1,4]

The electrons and ions in the plasma are completely free to move and can give rise to current. The electrons due to their light and fast nature are the main contributors, and ions being heavy and slow provide a neutralizing background. Although plasma is electrically neutral, it is electrically conducting; it is an electrically neutral conductor capable of interaction with electric and magnetic fields. The force of interaction between plasma and neutral gas is entirely different, and this is the basic difference between them. In a neutral gas, the type of interaction between particles is of Van der Waal's type, short range as well as strong, that is, the interactions are dominated by isolated, distinct two particle collisions (binary collisions), whereas in a plasma, this force is of coulomb type, that is, long range and weak at very large distances. The long-range Coulomb force is a factor, which determines the statistical properties of the medium. The typical particle interaction energies are small compared to the thermal energies of the particle. Thus, plasma is a collective but weakly coupled medium in which interaction energies are much smaller than thermal energies. For a charged particle gas to be called as plasma, collective effects are important. These collective rather than binary charged particle interactions lead to a wide variety of interesting phenomena in plasma like collective (Debye) shielding of individual charges, dielectric medium response to perturbations, oscillations at the "plasma" frequency, and wave propagation.[1,3]

Generally, the average kinetic energy of constituent particles (electrons, ions, and neutrals) of plasma is different. Electrons possess greater value for average kinetic energy, ions follows next, and neutrals with the least, that is, plasma is a mixture of different constituent particles at different temperatures.[1]

14.2 DEFINITION OF PLASMA

Plasma is usually said to be a gas of charged particles but this definition went wrong in many situations. Two basic properties that are necessary for plasma are the presence of freely moving charged particles and their number should be large. Though these conditions are necessary, they are not sufficient. Plasma consists of neutral particles along with charged particles, and their relative number should affect the features of plasma. The total numbers of charged particle (the negatively charged electrons and positively charged ions) are in such numbers to make the whole charge zero so that plasma is electrically neutral. Among many, one of the good definition for plasma stated as "plasma is a quasineutral gas of charged and neutral particles which exhibits collective behavior."[1,4]

14.2.1 QUASI-NEUTRALITY

Quasi-neutrality means the obvious variation in the neutrality of plasma on a large macroscopic scale. If we inspect plasma on a microscopic scale, the charges (electrons and ions) making up the plasma can bring about charged regions and electric fields. As electrons are mobile, plasma is a very good conductor of electricity, and any charges that develop are immediately neutralized, and plasmas can be treated as being electrically neutral in various cases.[6]

Plasma maintains perfect charge balance. If q_e and q_i, respectively, represent the charge of electron and charge of ion, n_i and n_i, respectively, represents the number density of electrons and ions, then $-q_e \times n_e = q_i \times n_i \pm \Delta n$, where $\Delta n = n_e - n_i$, and $\Delta n \ll n_e, n_i$.[5]

Factors, like temperature and density of plasma, decide the distance over which the apparent variation of quasi-neutrality in plasma. For example, the region of quasi-neutrality will be small, if the density of plasma is high because the number of negatively charged and positively charged particles is nearly the same at high density. "Debye length" refers to the distance over which the quasi-neutrality breaks (https://www.plasma-universe. com/Quasi-neutrality).

14.2.2 COLLECTIVE INTERACTION

The interaction force between two neutral particles at a distance "r" varies as r^{-6} or r^{-7} (Van der Waal's force) is of short range, but at touching distances, it becomes suddenly very strong. The straight and zigzag path in the case of neutral particle results from this sudden impact. This type of motion is called Brownian motion. In the case of plasma, each electron or ion can interact with its innumerable neighbors simultaneously and continuously due to the long-range Coulomb force; hence, the path of particles in plasma is entirely different with that of a neutral system.

As in the case of a neutral system (a gas system), we cannot divide the path of the charged particle into straight lines by connecting two successive points where a collision occurred. Electrons and ions produce fields, and all the particles of plasma are always in the electric field produced by the other particles. The magnitude and direction of this electric field are not static, it is changing continuously. Nevertheless, the average field due to the arbitrary fluctuations will be zero on averaging over a long time, but it does not mean that the average field is completely zero always and all over. The magnitude and direction of the velocity of a charged particle change constantly, even if, the internal microfield inside the plasma is small. The direction change in the motion of a charged particle occurs incessantly, not abruptly, as the intensity of microfield being small on the average and of long range. Hence, collisions, in this case, are smooth, not sudden or strong.

14.3   CRITERIA FOR PLASMA

All ionized gases cannot consider as a plasma state. There are certain criteria needed to satisfy for calling ionized gases as plasmas. These criteria can be understood by discussing Debye shielding, plasma parameter, and plasma frequency.

14.3.1   DEBYE SHIELDING

The ability to shield out the electric potential that is applied to plasma is a fundamental characteristic of plasma behavior. Consider that two oppositely charged balls are connected to a battery to put an electric field inside the plasma. The positively charged ball attracts negatively charged particles and negatively charged ball attracts positively charged particles, forming a cloud of charged particles around the balls. (The battery is large enough to maintain potential so that recombination will not occur.) The number of charges in the cloud and the ball should be the same if the plasmas were cold and no thermal motions were present. Perfect shielding occurs, and outside the clouds, no electric field would be present in the plasma. If there is a finite temperature, particles at the cloud edge (weak electric field) get enough thermal energy to escape from the electrostatic potential well. Shielding is incomplete and kT/e is the order of potential which leaks into the plasma causing the existence of a finite potential there.[16]

Interactions between individual charged particles are insignificant compared to collective effects. This introduces the concept of Debye screening or Debye shielding. The distance over which quasi-neutrality breaks is generally described by Debye length (or Debye sphere, which is a sphere of radius as Debye length) and varies according to the physical characteristics of the plasma. In plasmas, found in gas discharge tubes, the charged regions do not exceed a millimeter which means that the Debye length is typically less than a millimeter (https://www.plasma-universe. com/Quasi-neutrality).

The electrostatic potential of an isolated particle of charge "q" is $\phi = q/r$. In plasma, electrons are attracted to the vicinity of an ion and shield its electrostatic field from the rest of the plasma. Similarly, an electron at rest attracts ions and repels other electrons. This effect alters the potential in the vicinity of a charged particle. The potential of a charge at rest in plasma is

$$\phi = \frac{q}{r} e^{-r/\lambda_D}$$

where λ_D is the Debye length.[2]

The Debye length is given by $\lambda_D = \sqrt{\varepsilon_0 K_B T_e / e^2 n_e}$, where ε_0 is the permittivity of free space, K_B is the Boltzmann constant, T_e is the electron temperature, e is the charge of the electron, and n_e is the number density of electrons. It is a measure of the length over which fluctuating electric potentials that may appear in plasma correspond to the transition of thermal particle kinetic energy to electrostatic potential energy.[8,17]

The Debye length λ_D provides the measure of the distance over which the influence of the electric field of an individually charged particle is felt by the other charged particle. The Debye length is a measure of the sphere of influence of a given test charge in plasma. In general, the Debye length depends on the speed of the test charge with respect to the plasma. The Debye length is $(\lambda_D) \ll L$; this means that the extension of an ionized gas must be large compared to the Debye length to satisfy the condition of being plasma. Otherwise, there will not be sufficient space for the collective shielding effect to take place. The plasma charges effectively screen out the electric field of the test charge outside of the Debye sphere. This phenomenon is called Debye screening or shielding. The shielding of electrostatic fields is a consequence of collective behavior which exhibits between plasma particles to reduce the influence of externally introduced charge; the plasma particles act together in a coordinated way. This effect can only be observed if the Debye radius is much smaller than the size of the system, that is, $\lambda_D \ll L$. This condition is one of the necessary conditions for gas of charged particles to become plasma.[4,8,17]

14.3.2 PLASMA PARAMETER

Another criterion that is necessary to ensure that a gas of charged particles behave collectively and thus become plasma; it requires the number of electrons (N_D) inside the Debye sphere must be large $N_D \approx n_e \times \lambda_D^3$. The parameter $g = 1/N_D$ is called plasma parameter. For Debye shielding to occur, and for the description of plasma to be statistically meaningful, the number of particles in a Debye sphere must be large, that is, $g \ll 1$ (plasma approximation). This condition is necessary to ensure plasma criterion; in other words, the number of electrons, N_D, in the sphere of radius (λ_D) around the charge should $\gg 1$, which means that the average distance between electron must be very small compared to Debye length.[2,4,17] In plasma when many particles interact at the same time, the plasma parameter "g" is small. The plasma parameter can also be used as a measure of the ratio of the mean interparticle potential energy to the mean plasma kinetic energy. An ideal gas corresponds to zero potential energy between particles. If the plasma parameter is small, the plasma is treated as an ideal gas of charged particles, that is, a gas that can have a charge density and electric field, but in which no two discrete particles interact.[2]

To ensure that $n_e \lambda_D^3$ be large, the density must be low, since

$$g = \frac{1}{n_e \lambda_D^3} \frac{\alpha^{n_e^{1/2}}}{T^{3/2}}.$$

Because the collision frequency decreases with density n_e and temperature T, the condition g that tends to zero corresponds to a decreasing collision frequency. The dimensionless parameter "g" can be used as a measure of the degree to which plasma or collective effects dominate over single particle behavior. The plasma state can be described by equations obtained from an expansion of the exact many-body equations in powers of "g."[2]

14.3.3 PLASMA FREQUENCY

In two-component plasma, electrons are displaced from their equilibrium position; they experience a force to bring back them to the equilibrium position. The frequency of this simple periodic harmonic motion will be at plasma frequency. This phenomenon is referred to as plasma oscillation. It is a measure of the length of time required for an electron or ion moving with thermal speed to travel a Debye length and the electron density in plasma can also be specified using plasma frequency.

Because of the long-range forces exist between plasma particles, plasma behaves like a system of coupled oscillators. The angular frequency of collective electron oscillations called the (electron) plasma frequency, which is given by

$$\omega_{pe} = \sqrt{\frac{n_e e^2}{m_e \varepsilon_0}},$$

where m_e represents the mass of electron and plasma frequency represents the most fundamental time scale in plasma physics. The electrons, ions, and neutrals in plasma are at different plasma frequencies. However, electrons are relatively fast, by far, the most important, and "plasma frequency" refers to electron plasma frequency. The amplitude of this collective oscillation may be damped by the electron–neutral collision. The electron–neutral collision frequency (v_{en}) should be small when compared with electron plasma frequency $(v_{pe} = \omega_{pe}/2\pi)$ so that damping caused by electron–neutral collision will not diminish the amplitude of collective oscillation, that is, $v_{pe} > v_{en}$. The condition maintained for the gas of charged particles to behave like plasma is

$$\omega_p \times \tau > 1$$

where τ is the mean time an electron travels between the collisions with the neutral particle, and ω_p represents the angular frequency of typical plasma oscillation. The above condition signifies that on comparing with the typical time during the plasma physical parameters changing the mean time for the collision between neutral particle and electron should be large.[5,7,8,17]

14.4 TYPES OF COLLISIONS IN PLASMA

Depending upon the nature of plasma and other characteristics like mass, velocity, internal energy, and so on, different types of collisions may take place among the particles of the plasma, and these collisions can be broadly classified into two types: elastic and inelastic collisions.[1]

14.4.1 ELASTIC COLLISIONS

An elastic collision is a type of collision, in which the total kinetic energy of the interacting particles remains unchanged. Therefore, the internal energies of the particles also are the same. This type of collision can exist between two electrons, two ions, two neutral molecules between an ion and an electron, and between an electron and a molecule.

14.4.2 INELASTIC COLLISIONS

An inelastic collision is a type of collision, in which both the internal and kinetic energies of the interacting particles vary. All type of particles, like electrons, ions, atoms, molecules, ionized atoms or molecules, excited atoms or molecules, photons, and so on, can take part in this type of collision. Inelastic collisions involve excitation, ionization, recombination, charge transfer, attachment, dissociation, and so on.

Let A, B are neutral particles; A^* and A^* are excited particles; A^+ and B^+ are ions; and υ represents frequency of photon.

Excitation:

By photons $A + h\upsilon \rightarrow A^*$; $\quad A^* \rightarrow A + h\upsilon$ (spontaneous de-excitation), $A^* + h\upsilon \rightarrow A + 2h\upsilon$ (stimulated emission of a photon)

$$A + e^- \rightarrow A^* + e^- \text{ (by electrons); } A + B^+ \rightarrow A^* + B^+ \text{ (by ions);}$$

$$A + B \rightarrow A^* + B \text{ (by atoms)}; A + B^* \rightarrow A^* + B \text{ (by excited atoms)}$$

Ionization:

$$A + h\upsilon \rightarrow A^+ + e^- \text{ (by photons)}; A + e^- \rightarrow A^+ + e^- + e^- \text{ (by eletrons)};$$

$$A^* + e^- \rightarrow A^+ + e^- + e^- \text{ (by electrons)}; A + B^+ \rightarrow A^+ + B^+ + e^+ \text{ (by ions)};$$

$$A + B \rightarrow A^+ + B + e^- \text{ (by atoms)}; A + B^* \rightarrow A^+ + B + e^- \text{ (by excited atoms (penning effect))}$$

Recombination:

$$A^+ + e^- \rightarrow A + h\upsilon \text{ (ion} - \text{electron)}; A^+ + B^- \rightarrow AB + h\upsilon \text{(radiative)};$$

$$A^+ + B^- \rightarrow A^* + B^* \text{ (neutralizing)}; AB^+ + e^- \rightarrow A + B^* \text{ (dissociative)};$$

Charge transfer:

$$A^+ + B \rightarrow A + B^+; \quad A^{2+} + B \rightarrow A^+ + B^+; A^{2+} + B \rightarrow A + B^{2+}$$

Attachment:

$$A + e^- \rightarrow A^- + h\upsilon \text{ (radiative)}; A_2 + e^- \rightarrow A + A^- \text{ (dissociative)};$$

$$AB + e^- \rightarrow A^- + B \text{ (dissociative)}; A_2 + e^- \rightarrow A_2^- \text{ (molecular)};$$

Dissociation:

$$AB + h\upsilon \rightarrow A + B \text{ (photo-dissociation)}; A_2 + e^- \rightarrow A^+ + A^- + e^-;$$

$$AB + e^- \rightarrow A^+ + B^- + e^-$$

Along with these processes, other processes can also take place. The collision cross-section depends on the energies of the particles involved.

14.5   GAS DISCHARGE PLASMA

The term "discharge" refers to the flow of electric current through an ionized gas or it is the process of ionizing gas by the application of an electric field. An ordinary gas is nearly a perfect insulator, that is, electrical conductivity is very small; but in the plasma state, it becomes highly conducting. As gases are ionized to an adequate degree, it emits radiation. The gas discharge physics deals with the generation of electric currents in gases and sustaining the capability of gas to transmit electricity and absorb electromagnetic radiation.[9]

The gas discharge is an example of a self-sustaining system so that the electrons that depart the gas discharge plasma must be reproduced in the system by means of some ionization process. The self-sustainability for any type of gas discharge is due to the ionization balance. It represents a typical gas discharge tube; here, as a result of ion bombardment with a cathode, secondary electrons are formed at the cathode, and the multiplication of secondary electrons take place by means of ionization collision with atoms. This is referred to as self-maintenance of gas discharge. Two parameters, first Townsend coefficient (α) and second Townsend coefficient (γ) are important factors governing the self-maintenance character.[10] In addition with α and γ process, other processes involved are β effect, δ effect, ε effect, η effect, and σ effect. Ions produced by initial electrons, which causes ionization of gas, are referred to as β effect. The photons, which produced by means of recombination of electrons and ions in gas, on being bombarded with the cathode produces secondary electrons, and this effect is termed as δ effect. The excited metastable atoms on incidence with cathode produces secondary electrons, and this is called ε effect. η Effect refers to photoionization and σ effect refers to disturbance caused by the space charge on the electric field between the electrodes and the resultant creation of a spark.

Consider two metal electrodes (K—cathode and A—anode) connected to a DC power supply, which is inserted in a glass tube. The tube is evacuated and filled with a suitable gas (e.g., Argon). The pressure of the gas can be varied. The voltage between electrodes and current in the circuit is measured.

If a low voltage is applied to the electrodes (several 10s of volts), no visible effect is produced. The effect of ionization of gas emits light. Using the data, the V–I characteristics of gas discharge plasma can be plotted.[9]

The voltage–current (V–I) characteristics in a gas discharge plasma can be obtained by measuring the voltage between electrodes and current in the circuit. Current (I) taken along the x axis and voltage (V) along the y axis. In the typical nature of V–I characteristics of gas discharge, the nature of V–I characteristics of the discharge consists of three distinct sections, they are the dark discharge (Townsend discharge), glow discharge, and arc discharge. The gas used, pressure of the gas, electrode material, geometry of the electrodes, and interelectrode distance are the factors on which the breakdown voltage, the V–I characteristics, and the structure of discharge depends (https://www.plasma-universe.com/Electric_glow_discharge).

14.5.1 TOWNSEND DISCHARGE REGION (DARK DISCHARGE)

The region between A and E of V–I characteristics is termed as Townsend discharge. Except for corona discharge, the discharge is not visible to the eye (https://www.plasma-universe.com/Electric_glow_discharge). In this section A to B, current is proportional to voltage. In the second section

B to C, even if, the voltage continues to increase, the current remains almost constant. This is known as the saturated current region. In these two regions, the discharge is nonself-sustained means without an external agent, the discharge cannot continue. In part C to D, the current is found to increase with further increase in voltage, but then a region is established for which even if the voltage remains basically constant, the current continues to increase rapidly. This region is known as Townsend discharge. In effect, no glow is observed since the current through this part is extremely small (10^{-6} A). This part of the discharge is self-maintained.

In Townsend dark discharges, Corona discharges come in high electric field near sharp points, edges, or wires in gases prior to electrical break-down. Corona discharges are technically glow discharge which is visible to the eye if the coronal currents are essentially high. The whole corona region is dark for low currents, as suitable for the dark discharges.

As the electric field becomes stronger, electrical breakdown occurs in Townsend regime with the addition of secondary electrons emitted from the cathode. At breakdown potential (V_B), the current might be increased by a factor 10^4–10^8 and is restricted only by the internal resistance of the power supply connected between the electrodes. The discharge tube cannot draw enough current to breakdown the gas if the power supply has a very high internal resistance. If the internal resistance is low, then the breakdown of the gas will take place at the voltage (V_B) and go to the region called normal glow discharge (https://www.plasma-universe.com/Electric_glow_discharge).

14.5.2   GLOW DISCHARGE REGION

The plasma in this region is luminous and so it got the name glow discharge. Just beyond Townsend discharge, even if the voltage is reduced, the current is found to increase continuously. When dV/dI and hence the dynamic conductivity is negative, this region constitutes a metastable state. It is a transition zone between the Townsend discharge and glow discharge that immediately follows. The current continues to increase at a constant voltage in the glow discharge region, and the values of V and I are, respectively, about 10^2 V and 10^{-4}–10^{-2} A.

The gas enters the normal glow discharge region after a transition from E to F; here, over several orders of magnitude in the discharge current, the voltage is almost independent of the current. The electrode current density not depends on the total current in this region. The fraction of cathode occupied by the plasma increase as current is increased from F to G. At point G, plasma covers the complete cathode surface.

The region between G and H refers to the abnormal glow discharge region. Above the point G, the voltage increases considerably with the increasing total current. Hysteresis can be observed in the V–I characteristics on moving to the left from the point G. At appreciable lower currents as well as current densities than at point F, the discharge maintained here and then only makes a transition back to Townsend region.

14.5.3   ARC DISCHARGE REGION

In this region, the dynamical resistance and conductivity are negative, indicating another metastable state and transition region. The voltage reduces and reaches a steady value with increase in current, independent of current. The voltage remains steady at about 1 V; the current undergoes a very large change 10–10^3 A. With high current, but low voltage, this region is referred to as the arc discharge.

The electrode becomes adequately hot that the cathode emits electrons thermionically at the point H. The discharge will be subjected to a glow to arc transition, H–I, if the internal resistance of the DC power supply is sufficiently low. From I to J as the current increases, discharge voltage decreases. After the point J, for a large increase in current, the voltage increases very slowly (almost steady).

14.6   STRUCTURE OF GLOW DISCHARGE PLASMA

Glow discharge plasma is a self-sustaining discharge with a cold cathode emitting electrons due to secondary emission mostly due to positive ion bombardment.[9] It is a collision-dominated plasma, in which electron-neutral inelastic, exciting collisions playing the dominant role.[13] The two most important collisions take place in the glow discharge system are excitation and ionization of gas atoms. The excitation collisions and the subsequent deexcitations to lower levels by radiative decay are responsible for the characteristic glow in the glow discharge plasma.[11,12]

A discharge tube is used for the generation and analysis of the glow discharge. It is one of the most studied and widely applied types of gas discharge. The glow discharge pattern consists of dark and bright luminous layer. The pattern is easily observed at low pressures, the glow discharge pattern has an enchanting beauty. We can observe several layers between cathode and anode in the discharge tube.[9]

14.6.1   THE ASTON DARK SPACE

Electrons are ejected from the cathode at energies less than 1 eV, which is not enough for exciting an atom (15 eV is required for exciting Ar). This results in the formation of the Aston dark space. These electrons accelerate in the strong electric field near the cathode since the electron energy is too low and collisions with the neutrals do not lead to ionization. The Aston dark space has a negative space charge, which means that the electrons outnumber the positive ions in this region. Since there is no ionization collision take place, so it appears dark (less luminous). This region is a thin region close to the cathode (http://www.glow-discharge. com/?Physical_background:Glow_Discharges).[13]

14.6.2   THE CATHODE GLOW

This region comes next to the Aston dark space. The electrons accelerated by the field, and the energy of electrons is sufficient for exciting neutral atoms during collisions. Two or even three layers of cathode glow may be formed. They correspond to the excitation of different atomic levels, lower ones closer to the cathode and higher ones farther out. The cathode glow sometimes hides the Aston dark space as it comes very closer to the cathode. The ion density of the cathode glow is relatively high. The cathode is bombarded by positive ions coming mainly from the negative glow region, where they are created by collision with fast electrons. These ions recombine at the cathode and then fall to the ground state with the emission of light (http://www.glow-discharge. com/?Physical_background:Glow_Discharges).[13]

14.6.3   THE CATHODE DARK SPACE (COORKES DARK SPACE)

This region is a comparatively dark region, having a strong electric field, positive space charge with a relatively high ion density. The electrons are accelerated by the electric field in this region. The positive ions are accelerated to the cathode, which causes the emission of secondary electrons and pulverization of the cathode material. These secondary electrons will be accelerated, and through collision with neutrals, creation of new ions takes place. In this way, the continuity of the process is ensured and the glow discharge is self-sustaining. The large potential difference over a small distance gives rise to a high electric field in the cathode dark space. Almost the entire potential difference between anode and cathode falls off in the cathode dark space; therefore, the cathode dark space is also called "cathode fall" (V_c). Due to high cathode fall, the electrons will be accelerated to high velocities and they will not play a significant role in determining the space charge. Since ions are massive than electrons, the ions will not reach such high velocities and therefore ions determine the space charge. Hence, the cathode dark space is characterized by a highly positive space charge. The electrons which make ionizing collision in the cathode glow region are slowed down and the newly generated electrons have very small energy so, at the right side of the cathode glow, the electrons do not have sufficient energy to make ionization collision. Hence, this dark space is called the cathode dark space (http://www.glow-discharge.com/?Physical_background:Glow_Discharges).[12,13]

14.6.4   THE NEGATIVE GLOW

This region is the brightest intensity region of the whole discharge. This region is more or less equipotential and field free. In the cathode region, electrons that are accelerated to high speed produce ionization and inelastic collisions of slower electrons produce excitations. These slow electrons are the predominant reason for the generation of the negative glow region. The positive and negative space charges in this region are equal to each other, which results in the neutrality of charge. Even so, in the negative glow region electrons carry essentially the total current due to their high mobility. The region with the most numbers of exciting and ionization collisions occur in this region due to the high density of both positively and negatively charged particle. The negative glow is characterized by a bright light. The electrons lost most of their energy at the end of negative glow, excitation, and ionization no longer exist, resulting in much lower light intensity at the end region and the start of the new dark region.

14.6.5　THE FARADAY DARK SPACE

Faraday dark space separates the negative glow from the positive column. It is a dark and nearly equipotential region. The Faraday dark space can be considered as a repetition of Aston dark space. The Faraday dark space draws electrons that leave the negative glow with too low energy for excitation; therefore, the Faraday dark space is rather dark. The electric current in this region is, therefore, carried by the electrons.

14.6.6　THE POSITIVE COLUMN

The positive column of a DC glow discharge is the best pronounced and most widespread example of a weakly ionized nonequilibrium plasma sustained by an electric field.[9] Although it is the largest part of the entire discharge, the positive column is not necessary for maintaining the discharge. It is only present in the discharge at sufficiently large cathode–anode distances. The properties of this region are the closest to those of plasma; it is almost neutral overall with almost no electric field. The electric potential is almost constant throughout the length of the column, which means that there is no net space charge distribution and a very weak electric field. The electric field in this region is about 1 V/cm, just large enough to maintain the required degree of ionization. The electron number density is typically 10^{15}–10^{16} electrons/m^3. Electrons will again gain energy by the electric field. The interface between Faraday dark space and positive column is defined as the position where the electrons have enough energy for excitation and ionization. Therefore, the positive column is a bright zone. Sometimes, the positive column is not uniform and filled with striations (moving or stationary bright and dark layers) (http://www.glow-discharge.com/?Physical_background:Glow_Discharges).[12]

14.6.7   THE ANODE GLOW

This region is a bright region (slightly brighter than the positive column). The anode glow is not always observed. The intensity is low. The collision is small than in the positive column. This indicates that the energy of some particles has reduced, so they do not make ionization and hence the lower intense anodes glow.

14.6.8   THE ANODE DARK SPACE

The space between the anode glow and the anode itself is called the anode dark space or anode sheath. The anode attracts the electrons from the positive column and returns them to the external circuit. Due to electrons moving toward the anode, this region has a negative space net charge density. The electric field is comparably higher than that of the positive column. At this region, almost all the particles lose their energy, so we got anode dark spot.

14.7   APPLICATIONS OF DC GLOW DISCHARGE PLASMAS

Gas discharge and related plasmas have well-established applications in the field of material technology, microelectronic industry, surface treatment (etching of surfaces for IC fabrication), plasma polymerization, thin-film deposition, lasers, and so on. In the field of life sciences, atmospheric pressure plasmas have applications, sterilization of materials (bacteria killing), treatment of skin diseases, and so on. Fluorescent lamps, gas lasers, neon advertisement, plasma-display panels, and so on, use the excitation, and light-emitting characters of discharge plasma (https://www.uantwerpen.be/en/rg/plasmant/reseach/research-topics/gas-discharge-plasma/).

DC glow discharge has a wide range of applications for cleaning material surfaces, for pumping of gas discharge lasers, thin film deposition, and so on. Glow discharge plasma is a tool for sputtering, etching, activation of gaseous atoms, and molecules. In the fabrication of silicon-based IC, plasma etching has been widely using (https://www.uantwerpen.be/en/rg/plasmant/reseach/research-topics/gas-discharge-plasma/).[11,14]

14.7.1   GLOW DISCHARGE AND SPECTROMETRY

In a glow discharge tube, the ions inside the plasma accelerated in the electric field and move toward the cathode and release atoms of the cathode material, and this is called sputtering. These atoms enter the plasma and they also undergo collisions. This is the basis of using glow discharge for analytical applications. The analytically important element should be used as the cathode of glow discharges plasma. These cathode atoms, in turn, sputtered via ion bombardment and arrive into the plasma. Therefore, the plasma can be considered as an atom reservoir with compositional characteristics of the cathode (material to be analyzed is taken as cathode). Using an external light source, the atoms can be probed, measuring the resulting absorption or fluorescence. Hence, we can use glow discharges for atomic absorption spectrometry (GD-AAS) and atomic fluorescence spectrometry. The absolute number of densities of plasma species can be obtained from AAS. The sputtered atoms undergo collisions, leading to ionization of the element to be analyzed; the ions can be measured in a mass spectrometer, which leads to glow discharge mass spectrometry (GDMS). As an ion source in mass spectrometry, glow discharge was well known. A board range of sample types is allowed for analysis using GDMS. Using GDMS isotopic ratio measurements in metals can also be carried out. The excitation of the material to be measured, when deexcites emit characteristic photons; these can be detected by an optical emission spectrometer, this results in glow discharge optical emission spectrometry (GD-OES). GDMS and GD-OES comprise most of the applications of glow discharge spectrometry.[12]

14.7.2   SURFACE MODIFICATIONS

In microelectronics industry, for IC fabrication, one-third of the different steps are plasma-based oxygen discharges to grow SiO_2 films on silicon, discharges of CF_4 or Cl_2 or O_2 are used selectively remove Si films, O_2 discharges to remove photoresist or polymer films, and so on. These types of steps are repeatedly done in the fabrication of IC, different steps in making an IC.

The above six steps: (1) thin-film deposition on substrate; (2) photoresist film deposition over the film; (3) the resist selectively exposed to light through a pattern; (4) The resist is developed, removing the exposed resist regions, leaving behind a patterned resist mask; (5) etching is done on the film to obtain the pattern of the developed photoresist, the mask protects the underlying film from being etched; and (6) finally, reaming mask is removed. Generally, film deposition and etching are done by plasma processing.[11]

14.7.3 DEPOSITION OF THIN FILM

Sputtering is a phenomenon observed in glow discharge plasma at an adequately high voltage. In physical sputtering, from the plasma, ions, and atoms bombard the target and release atoms and molecules of the target material along with secondary electrons. At the substrate material, the sputtered atoms arrive by diffusion through the plasma and the sputtered atoms can deposit on the substrate. In reactive sputtering, a reactive gas is used. Along with positive ions, the dissociation products from the reactive gas also interact with the target. Therefore, a combination of both reactive gas and target material will deposit as film on substrate.[11]

14.7.4 ETCHING

As an alternative to wet chemical etching on chips, glow discharge plasmas are often using to generate topographical patterns on chips. It offers important advantages over wet chemical methods. It provides sharp etching, pollution free, finer resolution, dry, and safe. The etching experiment is performed by a conventional reactive ion etching technology. This technology uses chemically reactive plasma to remove material deposited on wafers. The wafers are positioned on the electrode surfaces. The wafer surface is attacked by the high-energy ions from plasma and reacts with it. The etchant gas flows at a constant rate. The DC power supply is switched on to start the etching for constant process time. Using an optical microscope, the etch profile can be examined.[14]

14.7.5 PLASMA POLYMERIZATION

Physical and chemical modifications at the surfaces of a polymer can be achieved when plasma is brought in contact with the polymer (changes in cross-linking or molecular weight, producing more reactive sites, etc.). By this way, desirable properties such as wettability, barrier protection, biocompatibility, and so on, can be achieved. Without changing the bulk properties, plasma surface treatments allow the modification of polymers to achieve improved bonding. Using plasma polymerization, besides of surface activation, deposition of thin polymer films can also be achieved. The deposition of polymer films through plasma dissociation and to the excitation of an organic monomer gas and subsequent deposition and polymerization of the excited species on the substrate surface, which is essentially plasma-enhanced chemical vapor deposition technique. The deposited films are called plasma polymers, and which are chemically as well as physically different from conventional polymers.[11]

14.7.6   CLEANING OF SURFACES

Conventional wet cleaning technique can cause problems on cleaning the narrow channels of IC's using liquids. Plasma-assisted cleaning technique can avoid this problem. Cleaning the surfaces of materials in semiconductor industry means the removal of all possible undesirable residues including organic, oxides, and metallic contaminants on the surfaces of materials. The volatilization and subsequent removal of the volatile components from the gas phase, if the contamination is organic. The surface contaminants can be converted to volatile oxides by oxidizing it with plasma. Thus, cleaning plasma contains an oxidizing gas. The oxygen gas is an example.[11]

14.7.7   ASHING

Ashing is closely related to cleaning. It is mainly used to remove organic fragments from inorganic surfaces. The organic fragments consist of carbon and hydrogen. They will not react with molecular oxygen at room temperature. The plasma used for this purpose is an oxygen plasma, because in plasma, the atomic oxygen created is highly reactive, and at room temperature, it easily reacts with the organic fragments, forming CO, CO_2, and water vapor (volatile products), which can be pumped out of the machine.[11]

14.7.8 TOKAMAK

Large efforts were forwarded by the world to make the dream fusion power a reality. Confined hot and dense plasma for a sufficiently long time is required for accomplishing economical fusion power. To confine the plasma effectively, many different configurations are proposed. Tokamak is the most successful among them, which can hold hot and dense plasma. Tokamak is a vessel which is torus-shaped, in which the toroidal field coils produce the toroidal magnetic field and the superposition of poloidal field coils, as well as a current flowing within the plasma, produces the poloidal magnetic field. Russia, in the late 1950s, invented the most successful fusion device tokamak. However, some of the most important developments in understanding the nature of transport, impurity control improvement, and starting new large-scale tokamaks, and so on, were done during the 1980s. The tokamak fusion test reactor at Princeton, NJ, USA, by producing about 10 MW of fusion power in 1994 established a new record.[18]

International thermonuclear experimental reactor (ITER), the world's first and largest fusion reactor, experimental tokamak reactor situated in France, leads the exploration of fusion technology. The seven members, which are contributing to ITER are the European Union, China, India, Japan, South Korea, the United States of America, and Russia. India joined ITER on December 5, 2005. ADITYA and Steady-State Superconducting Tokamak are the two operational tokamaks in India situated at Institute for Plasma Research, Gujarat.

To condition, the plasma vessel walls in tokamaks and other fusion devices, DC glow discharge is a common technique. In ITER, this is one of the primary conditioning techniques to control the surface state of the plasma-facing components. Inside a toroidal chamber, the fusion reaction in tokamak takes place. The fusion reaction should be stopped after a certain period of operation and using glow discharge plasma, the inner walls of toroid should be cleaned. The glow discharge plasma is a low-temperature plasma discharge which operated in the absence of a toroidal magnetic field; between anodes (can be more than one) inserted into the tokamak vessel, the entire vessel wall acted as a cathode. Impurities play a major role in the behavior of tokamaks. Low-temperature partially ionized hydrogen plasma can be used to reduce the oxygen contaminations on the walls by conversion to water vapor. In high-temperature discharges, the high electron temperature can cause the breakup of water molecule.[6,15,19]

KEYWORDS

- applied electric field
- ionization of gas
- mechanism
- glow discharge system
- plasma

REFERENCES

1. Goswami S. N. *Elements of Plasma Physics* (2011), New Central Book Agency (P) Limited, Kolkata, India.
2. Krall, N A.; Travelpiece, A. W. *Principles of Plasma Physics—International Series in Pure and Applied Physics*; McGraw-Hill: New York, 1973.
3. Callen, J. D. *Fundamentals of Plasma Physics*; University of Wisconsin: Madison, WI, 2003.
4. Gedalin, M. *Introduction to Plasma Physics*; Ben-Gurion University: Beer-Sheva, Israel, 2006.
5. Howard, J. *Introduction to Plasma Physics* (2002), Plasma Research Laboratory, Australian National University, Australia.
6. Hagelaar, G. J. M.; Kogut, D.; Douai, D.; Pitts, R. A. Physical Principle. *Plasma Phys. Ctrl. Fusion* **2015**, *57*, 025008 (14 pp).
7. Fitzpatrick, R. *Plasma Physics—An Introduction*; CRC Press: Boca Raton, FL, 2014.
8. Wiesemann, K. *A Short Introduction to Plasma Physics*; Aept: Ruhr-Universität Bochum: Germany, 2014.
9. Raizer, Y. P., *Gas Discharge Physics*; Springer-Verlag: Berlin, Heidelberg, 1991.
10. Smirnov, B. M. *Uspekhi Fizicheskikh Nauk* **2009**, *52* (6), 559–571.
11. Bogaerts, A.; Neyis, E.; Gijbels, R.; Van Der Mullen, J. (2002), *Spectrochimica Acta* Part B 57.
12. Bogaerts, A.; Gijbels, R. (1998), *Spectrochimica Acta* Part B 53, 1–42.
13. Plasma Science and Fusion Energy Institute. *DC Glow Discharge*, Journal Published by Plasma Science and Fusion Energy Institute, Princeton Plasma Physics Laboratory, July 22–August 2, 2002.
14. Chiad, B. T.; Al-Zubaydi, T. L.; Khalaf, M. K.; Khudiar, A. I. *Indian J. Pure Appl. Phy.* **2010**, 48.
15. Oren, L.; Taylor, R. J. *Nucl. Fusion* **1977**, (6).

16. Chen, F. F. *Introduction to Plasma Physics and Controlled Fusion*; Springer Private Limited: India, 2006.

17. Bittencourt J. A. *Fundamentals of Plasma Physics*; Springer-Verlag: New York, 2004.

18. Deshpande, S.; Kaw, P. *Sadhana* **2013,** *38* (5), 839–848.

19. Fu, Y.; Luo, H.; Zou, X.; Wang X. *Plasma Sources Sci. Technol.* **2014,** *23* (065035), 1–6.

CHAPTER 15

INVESTIGATION OF WOOD SURFACES MODIFICATION BY RADIO-FREQUENCY PLASMA

IGOR NOVÁK[1], ANGELA KLEINOVÁ[1], JÁN MATYAŠOVSKÝ[2], PETER JURKOVIČ[2], PETER DUCHOVIČ[2], KATARÍNA VALACHOVÁ[3], and LADISLAV ŠOLTES[3]

[1]Department of Composite Materials, Polymer Institute of the Slovak Academy of Sciences, SK 84541 Bratislava, Slovakia

[2]Polymers and Testing Department, VIPO, Gen. Svobodu 1069/4, SK 958 01 Partizánske, Slovakia

[3]Department of Cellular Pharmacology and Developmental Toxicology, Centre of Experimental Medicine, Institute of Experimental Pharmacology and Toxicology, Slovak Academy of Sciences, SK 84104 Bratislava, Slovakia

*Corresponding author. E-mail: igor.novak@savba.sk

ABSTRACT

Various wood species, namely, oak, beech, maple, and ash modified by radio-frequency discharge (RFD) plasma in air atmosphere were studied. The physical and chemical changes were investigated by static contact angles of water and Fourier-transform infrared spectroscopy with attenuated total reflectance (FTIR-ATR) spectra measurements for all investigated woods. The results of the investigation confirmed the increase in the wood hydrophilicity/polarity in all cases, which is caused by the increase of –OH group concentration due to irradiation by RFD plasma. The content of COOH, C–O, and C=O groups after treatment by RFD plasma significantly increased and the water contact angles were diminished. FTIR-ATR spectroscopy confirmed that RF-plasma modifications of wood on the surface of all investigated kinds of wood samples lead to some changes that are also dependent on time of plasma exposure.

15.1   INTRODUCTION

Radio-frequency discharge (RFD) plasma can be suggested as the appropriate procedure for the hydrophilization of the polymeric surface. Due to the plasma treatment, the surface energy of wood is increased as a result of introduction of polar functional groups on the treated surface, thus making the surface of wood more hydrophilic.[1] RF discharge plasma at reduced pressure is currently an efficient method for treatment of surface and adhesive properties of wood, and it is considered as the "green" ecologically friendly method. For a wide industrial utilization, various woods have to possess a large set of various surface characteristics, including polarity, dye-ability, scratch resistance, tailored adhesive properties, antibacterial resistance, and so on.[1,2,13] The nanoscale dimension changes in the plasma-treated wood have been carried out, while maintaining the desirable material properties. The enhancement of wood hydrophilicity is a necessary condition to promote a better adhesion with water-based adhesives and coatings, which is currently being studied.[4,8,10,12]

There are two reasons why in the case of wood, discharge plasma could be applied for surface modification.[10] Firstly, discharge plasma in air itself significantly increases the hydrophilicity of wood because various polar groups are formed (e.g., hydroxyl, carbonyl, carboxyl, etc.), and the wood macromolecules are also cross-linked (up to a few microns), which leads to the increase in scratch resistance and to the improvement in barrier properties of the wood material. The second reason for the plasma use is an increase of adhesion in adhesive joint between polymeric adhesive and wood substrate growth of wood wettability, which is important for industrial applications. The low-temperature plasma represents a mixture of various excited particles, that is, ions, atoms, electrons, and radicals with a low degree of ionization and a little penetrating energy, but plasma particles have sufficient levels of energy to break chemical bonds on the wood substrate.[11] The treatment of wood by discharge plasma is limited to a 100 nanometers without affecting the bulk properties of the wood.[10] The increased surface polarity due to oxidation reaction during modification of wood by RF plasma improves its wettability and hydrophilicity.[3,9] The wettability assists in establishing a molecular-scale contact with the wood surface and is critical to the development of strong adhesion at the adhesive-wood interface.[7]

15.2   MATERIALS AND METHODS

The physical and chemical changes were observed using measurements of water contact angles by contact angle meter and FTIR–ATR for all investigated wood species.

15.2.1   MATERIALS

The samples of wood species such as oak, beech, maple, and ash with dimensions 50 × 15 × 5 mm (Technical University in Zvolen, Slovakia) and the moisture content of 8% were pretreated using RFD plasma in the air at the 100 Pa pressure.

15.2.1.1   RF-PLASMA MODIFICATION

The surface of wood was modified by RFD plasma. The modification of wood by the capacitive-coupled RFD plasma was performed in a RFD plasma reactor (Fig. 15.1) working at reduced pressure 80 Pa and composed of two 240 mm brass parallel circular electrodes with symmetrical arrangement, 10 mm thick, between which RFD plasma is created.

Two electrodes of RFD plasma reactor are placed in a locked-up stainless steel vacuum cylinder. One of them is powered and the other one is grounded together with steel cylinder. The voltage of RFD plasma reactor is 2 kV, frequency 13.56 MHz, current intensity was maximum of 0.6 mA, and the maximal power of the RFD plasma source is 1200 W. The wood samples were modified by RFD plasma at the power 350 W.

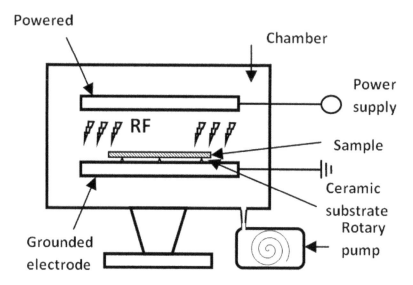

FIGURE 15.1   Scheme of RFD plasma reactor.

15.2.2   MEASUREMENT METHODS

The surface free energy of wood was measured by determination of contact angles (θ) with re-distilled water as the testing liquid.[10] The drops of testing liquid (V = 20 μL) were placed on the wood surface with a micropipette (Biohit, Finland), and the dependence θ = f(t) was extrapolated to t = 0. The contact angles measurements of water were measured using professional surface energy evaluation (SEE) system device completed with a web camera (Advex, Czech Republic) and necessary PC software. The measurements of contact angles were repeated 12 times and the arithmetic mean with measurement deviations has been taken into account.

Fourier-transform infrared spectroscopy with attenuated total reflectance (FTIR-ATR) measurements were performed with an FTIR NICOLET 8700 spectrometer (Thermo Scientific, UK) using a single-bounce ATR accessory equipped with a Ge crystal. For each measurement, the spectral resolution was 2 cm^{-1} and 64 scans were performed.

15.3   RESULTS AND DISCUSSION

The contact angle of water in the investigated wood surfaces diminished with the time of modification by RFD plasma (Fig. 15.2), and showed a steep decrease from 75° (pristine beech wood) to 40° after activation for 60 s of the studied sorts of wood by RFD plasma in air. As seen in Figure 15.2, the decrease in the contact angle of water can be explained by growth of the hydrophilicity of the investigated sorts of wood surface during pretreatment by RFD plasma in air. The hydrophilicity of the wood surface depends on the formation of polar oxygenic functional groups during RFD plasma modification of wood in air. After saturation (from 60 s of the plasma treatment) of the wood surface with polar groups, the hydrophilicity was stabilized. The efficiency of modification of wood by RFD plasma was lower in the case of ash. For ash, the dependence of water contact angle was lower than for other investigated wood species, that is, oak, beech, and maple wood.

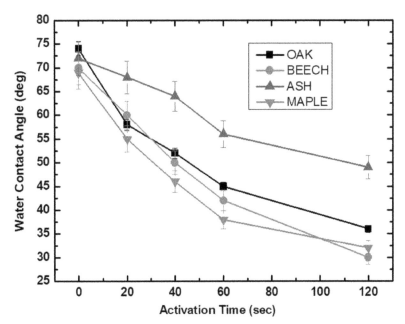

FIGURE 15.2   (See color insert.) Water contact angle of RFD plasma-treated wood species vs. plasma activation time.

The aging of RFD plasma-treated wood species is illustrated in Figure 15.3. The water contact angle of RFD plasma-modified wood during approximately the first 2 days after modification by RFD plasma increased faster, and after this period of time, the aging was slow. The increase in water contact angle during aging was lower in the case of ash than for oak, beech, and maple wood. The water contact angle of plasma-treated beech wood increased after 28 days of aging from 37° to 68°, that is, the growth by 45.6% was observed.

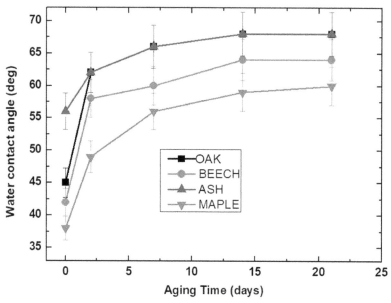

FIGURE 15.3 **(See color insert.)** Water contact angle of RFD plasma-treated wood species vs. plasma activation time.

In general, the spectrum of any type of wood is a mixed spectrum (composition) of cellulose and lignin with characteristic peaks in –OH bonds (with a maximum at about 3400 cm^{-1}) so as in the area of finger-prints, which is reflected in C–O–C, COO, and CH$_2$ bonds typical for polysaccharides. Moreover, the peak at 898 cm^{-1} glycosidic linkages, (C–O–C) appears for each of them as typical representative.[3] The so-called normalized spectra (Figs. 15.4–15.7 and Table 15.1), that is, the FTIR spectra modified by multiplying with a selected factor aiming to have the common Y axis for a better readability are in figures above. Thus, small changes in shapes of lines of C–H bonds and C–O–C, which confirms that changes in surfaces of the samples are highly visible. However, since it is the peak composed of more pieces, a different procedure was chosen to quantify these changes. Ratios of integrated intensities of oxygen bonding groups (with the majority contribution of OH groups) with their maximum at 3400 cm^{-1} and integrated intensities at the 2985 cm^{-1} $(CH_2)_{sym}$ were determined. However, this is only semiquantitative information, Table 15.1 shows ratios of integrated intensities P (–OH)/P (–CH$_2$), where the vibration of –CH$_2$– was chosen as an internal standard with the assumption that the plasma treatment does not affect this area.

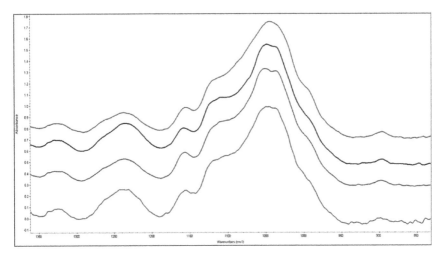

FIGURE 15.4 **(See color insert.)** FTIR spectra of oak wood treated by RFD plasma: the red color—untreated oak wood, the green color—plasma-treated for 20 s, the blue color—plasma-treated for 60 s, and the purple color—plasma-treated for 120 s.

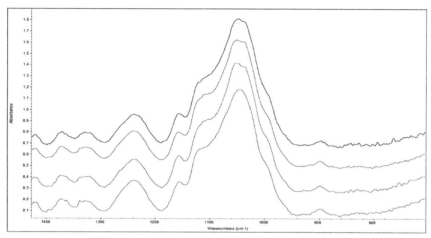

FIGURE 15.5   (See color insert.) FTIR spectra of beech wood treated by RFD plasma: the red color—untreated beech wood, the green color—plasma-treated for 20 s, the purple color—plasma-treated for 60 s, and the blue color—plasma-treated for 120 s.

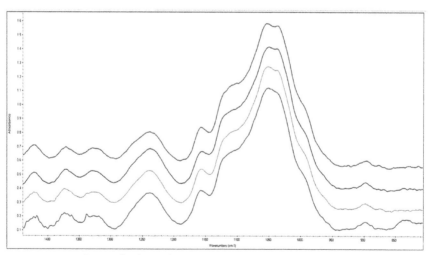

FIGURE 15.6   (See color insert.) FTIR spectra of maple wood treated by RFD plasma: the red color—untreated maple, the yellow color—plasma-treated for 20 s, the green color—plasma-treated for 60 s, and the purple color—plasma-treated for 120 s.

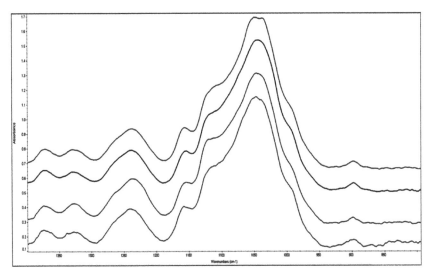

FIGURE 15.7 **(See color insert.)** FTIR of ash wood treated by RFD plasma: the red color—untreated ash, the green color—plasma-treated for 20 s, the blue color—plasma-treated for 60 s, and the purple color—plasma-treated for 120 s.

Table 5.1 shows the increased ratio of specified intensities for all kinds of wood compared to untreated samples. But just for one type of wood (maple), it may be declared that this ratio correlates with the time of exposure to plasma. These values are independent of time of the plasma treatment for oak; even in this case, the values for the treated samples are slightly higher than for untreated wood as well. The largest increase was observed in the case of beech wood (from value of 7.433 for the untreated wood up to 22.986 for 120 s of plasma treatment). The ratio $P(OH)/P(CH_2)$ for untreated types of woods ranges from 5.948 to 7.433, which indicates approximately the same hydrophilicity, that is, the content of –OH groups on the surface of all types of wood before any treatment was the same. The value of hardness given for each types of wood was determined according the Brinell hardness test. Values of ratios for ash with the hardness 4.0 are equal for all time regimes, which might be due to the fact that the ash is the hardest of all evaluated woods. Thus, the plasma treatment in this case is not as effective. Even the baseline of the ratio observed is the lowest just for ash. But it needs to be mentioned that the hardness values of the material are not dependent on the type of surface, in this case they depend on the porosity which may be the reason that the plasma treatment is more effective in some cases and less in the others. It can be concluded that FTIR-ATR spectroscopy confirmed that the plasma treatment causes some changes at the surface of all types of woods, which are dependent on the time of exposition in the case of maple. This trend is not completely proven for others species of wood, but it can be said that there is an increased content of hydrophilic groups compared to the untreated samples. Conclusions from the FTIR-ATR spectroscopy could be complemented by measuring of contact angles as well.

TABLE 15.1 The Ratio of Integrated Peaks for P (2895, CH_2 stretch) and P (3400, OH stretch) Calculated from FTIR for Wood Sorts (Oak, Beech, Maple, and Ash) Modified by RFD Plasma.

Oak, hardness = 3.7	$P(OH)/P(CH_2)$
Untreated	6.863
20 s plasma	9.549
60 s plasma	12.469
120 s plasma	17.414
Beech, hardness = 3.8	
Untreated	7.433
20 s plasma	13.973
60 s plasma	20.624
120 s plasma	22.986
Maple, hardness = 3.0	
Untreated	5.948
20 s plasma	11.414
60 s plasma	14.046
120 s plasma	16.960
Ash, hardness = 4.0	
Untreated	6.823
20 s plasma	10.841
60 s plasma	13.413
120 s plasma	13.243

15.4 CONCLUSIONS

FTIR-ATR spectra confirm the increase in the selected species wood polarity during the RFD plasma treatment in the air due to growth in the amount of –OH groups. The concentration of oxygen in the investigated wood after RFD plasma treatment increased. The amount of carbon during the plasma treatment of wood conversely decreased. The concentration of COOH, C–O, and C=O groups after treatment by RF-plasma significantly increased. Based on the FTIR results, it can be stated that there is an increased content of hydrophilic groups in wood compared to the untreated samples. The water contact angle of wood treated by RFD plasma in air decreased with activation time from 75° to 40°. The growth of water contact angle of plasma-treated wood from 37° to 68° during aging was faster within 2 days after the plasma pre-treatment. The water contact angles of four wood species showed a steep decrease after activation by RFD plasma in air. Water contact angles were markedly increased during the first 2 days of aging. RFD plasma-treated wood surfaces should be treated with procedures such as bonding, painting, and so on, up to 2 days after modification by plasma.

ACKNOWLEDGMENTS

This contribution was supported by the Slovak Research and Development Agency, projects, APPV-14-0506, APPV-16-0177, APPV-17-0456, and VEGA 2/0019/19.

KEYWORDS

- **Fourier-transform infrared**
- **contact angle**
- **hydrophilicity**
- **plasma treatment**
- **surface properties**
- **wood**

REFERENCES

1. Acda, M. N.; Devera, E. E.; Cabangon, R. J.; Ramos, H. J. Effects of Plasma Modification on Adhesion Properties of Wood. *Int. J. Adhes. Adhes.* **2012,** *32,* 70–75.

2. Bente, M.; Avramidis, G.; Förster, S.; Rohwer, E. G.; Viöl, W. Wood Surface Modification in Dielectric Barrier Discharges at Atmospheric Pressure for Creating Water Repellent Characteristics. *Holz als Roh- und Werkstoff* **2004,** *62,* 157–163.

3. Ciolacu, D.; Ciolacu, F.; Popa, V. I. Amorphous Cellulose: Structure and Characterization. *Cell. Chem. Technol.* **2011,** *45,* 13–21.

4. Frihart, C. R. Wood Adhesion and Adhesives (Chapter 9). In *Handbook of Wood Chemistry and Wood Composites;* CRC Press: London, Washington, 2005; p 504.

5. Kamdem, D. P.; Pizzi, A.; Triboulot, M. C. Heat-treated Timber: Potentially Toxic Byproducts Presence and Extent of Wood Cell Wall Degradation. *Holz als Roh- und Werkstoff* **2000,** *58,* 253–257.

6. Kleinová, A. High-density Polyethylene Functionalized by Cold Plasma and Silanes. *Vacuum* **2012,** *86,* 2089–2094.

7. Kúdela, J.; Štrbová, M.; Jaš, F. Influence of Accelerated Ageing on Morphology and Wetting of Wood Surface Treated with a Modified Water-based Coating System. *Acta Facultatis Xylologiae Zvolen* **2017,** *59* (1), 27–39. DOI: 10.17423/afx.2017.59.1.03.

8. Moghadamzadeh, H.; Rahimi, H.; Asadollahzadeh, M.; Hemmati, A. R. Surface Treatment of Wood Polymer Composites for Adhesive Bonding. *Int. J. Adhes. Adhes.* **2011,** *31,* 816–821.

9. Müller, G.; Schöpper, C.; Vos, H.; Kharazipour, A.; Polle, A. FTIR-ATR Spectroscopic Analyses of Changes in Wood Properties During Particle-and Fibreboard Production of Hard and Softwood Trees. *Bioresources* **2005,** *4,* 49–71.

10. Novák, I.; Popelka, A.; Krupa, I.; Chodák, I.; Janigová, I.; Nedelčev, T.; Špírková, M.; Odrášková, M.; Ráhel', J.; Zahoranová, A.; Tiňo, R.; Černák, M. Plasma Activation of Wood Surface by Diffuse Coplanar Surface Barrier Discharge. *Plasma Chem. Plasma Pro.* **2008,** *28,* 203–211.

11. Olaru, N.; Olaru, L.; Cobiliac, G. H. Plasma-modified Wood Fibers as Fillers in Polymeric Materials. *Romanian J. Phys.* **2005,** *50,* 1095–1101.

12. Reinprecht, L.; Šomšák, M. Effect of Plasma and UV-additives in Transparent Coatings on the Colour Stability of Spruce (*Picea abies*) Wood at Its Weathering in Xenotest. *Acta Facultatis Xylologiae Zvolen* **2015,** *57* (2), 49–59. DOI: 10.17423/afx.2015.57.2.05.

13. Wolkenhauer, A.; Avramidis, G.; Hauswald, E.; Militz, H.; Viöl, W. Sanding vs. Plasma Treatment of Aged Wood: A Comparison with Respect to Surface Energy. *Int. J. Adhes. Adhes.* **2000,** *29,* 18–22.

INDEX

Milton Keynes UK
Ingram Content Group UK Ltd.
UKHW050257161024
449569UK00042B/1751